BIOMEDICAL SENSORS

BIOMEDICAL SENSORS

Fundamentals and Applications

by

Harry N. Norton

Jet Propulsion Laboratory
California Institute of Technology

Library of Congress Catalog Card Number: 81-19022
ISBN: 0-8155-0890-5
Printed in the United States

Published in the United States of America by
Noyes Publications
Mill Road, Park Ridge, New Jersey 07656

Library of Congress Cataloging in Publication Data

Norton, Harry N.
Biomedical sensors, fundamentals and applications.

Bibliography: p.
Includes index.
1. Transducers, Biomedical. 2. Medical electronics. 3. Human physiology--Measurement. I. Title. [DNLM: 1. Biomedical engineering--Instrumentation. 2. Electronics, Medical--Instrumentation. QT 34 N884b]
R857.T7N67 610'.28 81-19022
ISBN 0-8155-0890-5 AACR2

Preface

The main objective of this book is to make a three-way introduction: to introduce those engaged in the medical field and those working in the electronics field to each other, and to introduce workers in both of these disciplines to electronic sensing devices, in general, and to biomedical sensors and the systems in which they are used, in particular.

Engineers and technicians in the specialized field of biomedical electronics will, no doubt, find useful information in this book. The real purpose of the book, however, is to communicate the basics of biomedical sensors and their use to people at all levels of skill: to medical and engineering students; to nurses and technicians; to service and maintenance personnel; to engineers, especially those entering the field of biomedical engineering; to doctors and surgeons; and to all those engaged in medical as well as veterinary research.

The book is structured to meet the "three-way introduction" objective. The first chapter presents a cursory overview of sensors, in general; it explains, very briefly, the systems they are used in, the many types of quantities that are commonly measured by sensors, their basic operating principles, and their salient characteristics. The chapter is written so that it should be able to be understood by anyone having a basic science background, without ever having received any background in the area of electronic sensing devices.

The second chapter provides a brief overview of physiology and anatomy. It is aimed primarily at those who have not been exposed to these subjects previously. Hence, workers in the medical field should be quite familiar with the subject matter presented; however, the emphasis is somewhat different from that usually noted in medical textbooks in that the systems of the body are described from the viewpoint of a measurement system engineer. This places the emphasis on sources of signals and measurements and on areas in which sensors are used.

The subsequent chapters, which really comprise the main body of the book, follow the organization by systems of the body and describe the sensing devices used in these systems. The last two chapters cover some commonly-used sensing and other diagnostic equipment and systems that are not body-system-specific, as well as electrical safety considerations.

A bibliography, at the end of the book, lists significant literature from which further information about sensors, in general, and biomedical sensors, in particular, can be obtained.

By meeting the objectives in this manner, it is sincerely hoped that the book will also meet its stated purpose.

Harry N. Norton

Acknowledgements

Leslie Cromwell and Fred J. Weibell contributed to this book by reviewing Chapters 2 through 8 and giving me the benefit of their comments.

Harry N. Norton

About the Author

Harry N. Norton is a member of the technical staff of the Jet Propulsion Laboratory, California Institute of Technology, Pasadena, California. He is a Fellow of the Instrument Society of America, and an Associate Fellow of the American Institute of Aeronautics and Astronautics. From 1964 to 1975, he was Director of Transducer Standards for the Instrument Society of America. He has lectured at and chaired numerous workshops in the field of sensors and transducers in both Europe and the United States. He is also the author of three other books dealing with transducers and sensors.

Table of Contents

1. Electronic Sensor Fundamentals

1.1 SENSORS

A *sensor* is an instrument which senses a specific quantity, property or condition (e.g., temperature, pressure, sound, strain) and provides an electrical signal in proportion to what it senses. More precisely, a sensor is a device which responds to changes in whatever it is intended to measure (changes in the *measurand*) with corresponding changes in its *output* (output signal). The relationship of measurand-vs-output of a sensor is a known relationship and is represented by the *calibration* (curve or table) of the sensor. Hence, the magnitude of the output of a sensor can be used to inform us about the magnitude of the measurand.

Most sensors provide an *analog output,* an output which varies continuously as the measurand varies. A few sensor types provide an inherently *digital output,* one which is quantized and coded; they are rarely used in biomedical applications; however, modern sensing equipment can incorporate an *analog-to-digital converter* which allows the signal to be displayed in digital form or fed into a digital computer. Some sensors are of the *switch type.* They produce an output signal when the measurand rises (or falls) to a preset level. It is this latter type of sensor that most of us are familiar with in daily life. Most homes, for example, use a thermostat to control a furnace. A thermostat is a switch-type temperature sensor, typically containing a bimetallic strip which bends with temperature variations. It is adjusted so that the amount of bending corresponding to a preselected temperature causes a pair of switch contacts to close. This completes the circuit to the furnace control and the furnace is turned on.

A similar device is used to illuminate a warning light in an automobile when the radiator temperature exceeds a certain limit. In some cars, however, an analog-output temperature sensor is used to indicate coolant temperature on a small meter. Similarly, a switch-type, or analog-output pressure sensor is used in the lubricant system of a car to illuminate a low-pressure warning light or provide an indication of oil pressure on a meter, respectively. The speedometer and (in some cars) the tachometer are other examples of displays of analog-output sensors in the automotive field.

Another familiar use of sensors is found in photographic light meters. A light sensor provides an analog output proportional to illumination and this output is displayed on an electric meter. By means of a chart or direct readings shown on the meter the camera can then be adjusted for the proper exposure. In many modern cameras a built-in light sensor provides the signal used for automatic exposure control. A comparison between the latter two applications illustrates how the same type of sensor can be used for providing information *(measurement)* or for control *(automatic control).*

Sensors are widely used in all sciences and industries. They are used for measurement and control of fluid-mechanical quantities (e.g., pressure, flow, moisture), solid-mechanical quantities (e.g., motion, force, strain, vibration), thermal quantities (temperature, heat flux), acoustic quantities (sound pressure, sound level), optical quantities (light intensity, color, refractive index), nuclear radiation (alpha particles, beta-, gamma- and X-rays, neutrons), electrical quantities (voltage, current, power, charge), magnetic flux density, and chemical properties and composition.

Many types of sensors are also called *transducers.* All transducers are sensors and the definition of "transducer" is the same as for "sensor." However, due to common usage of these terms, not all sensors are called "transducers." For example, in electro-optics, the term "light transducer" or "infrared transducer" is never used; instead, light sensors are very commonly called "light detectors" or "photodetectors." The same usage is found in nuclear radiation sensing where, e.g., gamma-ray sensors are commonly referred to as "gamma-ray detectors." Sensors for some specific measurands have also acquired unique nomenclature. Acceleration sensors are usually called "accelerometers," flow-rate sensors are called "flowmeters," strain sensors are called "strain gages" and force transducers (force sensors) are often called "load cells." Sensors of chemical properties and composition are generally called "analyzers," although the term "analyzer" really applies to the measuring system in which the sensor is used.

1.2 MEASURING SYSTEMS

Sensors are used in measuring systems of varying complexity. In all systems, however, the information needed by the observer is provided by a *display device.* Some devices provide a temporary display of the reading; electric meters (analog or digital) and oscilloscopes (cathode-ray tubes, CRTs) provide such displays. Other displays provide a permanent record; such devices include strip-chart records and various types of printers. Oscilloscopes and strip-chart recorders (oscillographs) are particularly useful when rapid fluctuations of a measurand must be observed. Meters and printers are more suitable when the measurand varies slowly. Discrete levels (e.g., "too high," "too low," "OK") can be displayed by a simple indicator light.

Figure 1-1 illustrates the most elementary measuring system. It consists of: a *sensor* which converts the measurand into an electrical output; *signal-conditioning* circuitry which modifies the sensor output into the kind of signal needed for proper operation of the display device; the *display device* which provides information about the measurand in the form desired by the observer; and a *power supply* which supplies the required electrical power to the signal conditioning circuitry, and which may also (as shown by dotted lines) supply *excitation power* to the sensor and may, additionally, furnish electrical power to the display device.

The power supply takes its power from a *power source,* usually either the available ac power in the building or a battery. Excitation power is needed by all sensors except the self-generating types (as explained in 1.4, below). The signal conditioning circuitry may contain a measuring circuit (e.g., a Wheatstone-bridge circuit for a resistive sensor), a simple dc amplifier, a dc-ac-dc converter, or other types of circuits or circuit combinations. A typical example of such a basic measuring system is the electronic thermometer.

Telemetry systems are used when the distance between the point(s) of measurement and the display is relatively long or when it is undesirable to use cabling between the sensor(s) and the display. The word is derived from the Greek *tele* (far, distant) and *metrein* (to measure). The link between the point of measurement and the point of display can either be via radio waves *(radio telemetry)* or via a transmission line *(hardline telemetry).* Radio telemetry is typically used in situations where signals from an ambulatory patient must be monitored on a central display console. Hardline telemetry is often implemented by using equipment that connects the sensing portion of the measuring system to the display portion by means of the telephone network.

In a basic telemetry system (Figure 1-2) the sensor output is condi-

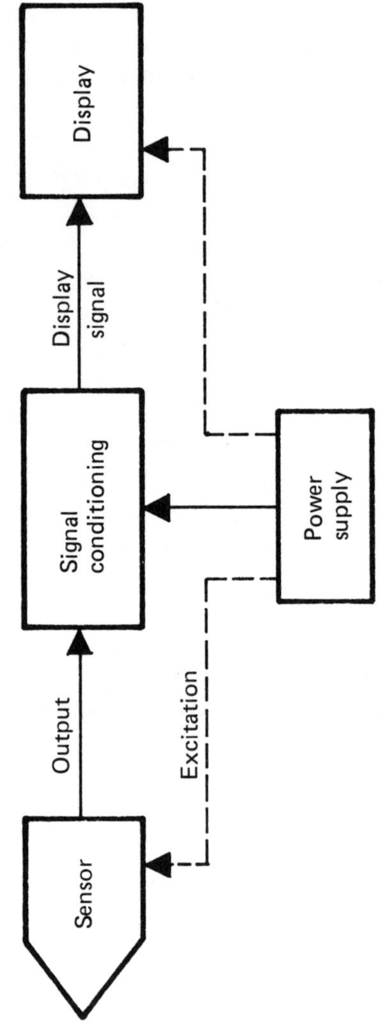

Figure 1-1. Basic measuring system

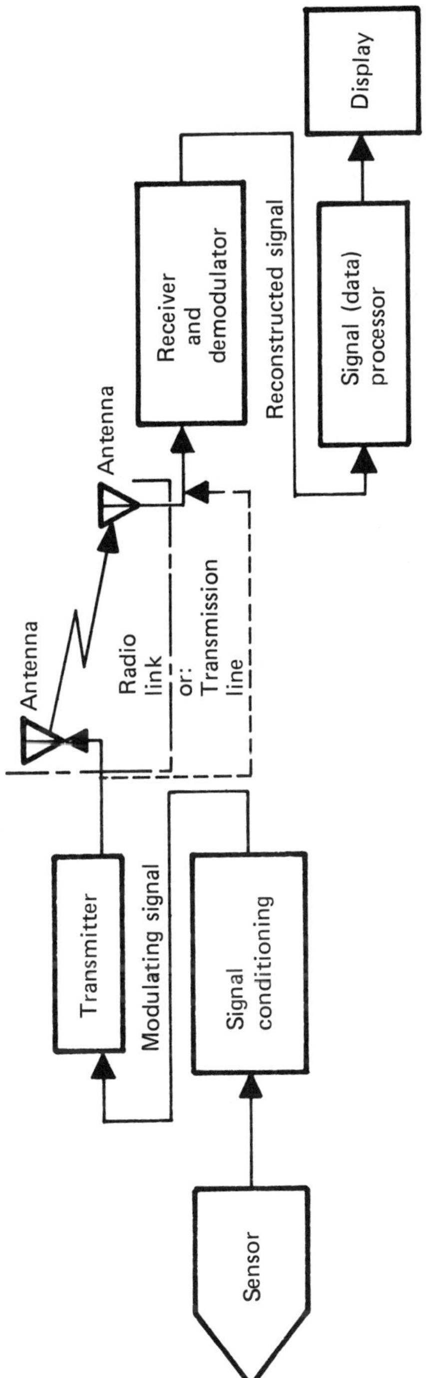

Figure 1-2. Basic telemetry system

tioned so that it is suitable for modulating a carrier signal. When a radio link is used the carrier frequency is in the high-frequency (radio frequency, RF) band. When the link is a transmission line it is either in the low-frequency portion of the RF band or (especially when telephone lines are used) in the audio frequency band. The *transmitter* amplifies the modulated carrier and couples this signal either to an antenna or to the transmission line (sometimes using ancillary equipment associated with the transmission line). The *receiver* gets the modulated carrier signal either from an antenna (in radio telemetry) or from a decoupling device connected to the transmission line. The *demodulator* portion of the receiver strips the modulation from the carrier so that signals equivalent to those at the output of the sensor signal conditioner are now reconstructed. The reconstructed signals are then conditioned (processed) into a form that the display device will accept.

The system illustrated in Figure 1-2 applies to *analog telemetry*. The essential element that must be added to obtain *digital telemetry* is an analog-to-digital converter, typically located at the output of the sensor signal conditioner. The modulating signal will then be in digital form. When signals from two or more sensors are to be handled by one telemetry system, the conditioned sensor outputs can be *multiplexed* (as explained for Figure 1-3, below) and are then synchronously demultiplexed, after demodulation, on the receiving end of the system.

Most modern measuring systems, particularly multisensor measuring systems, use digital techniques for signal processing and display and are then referred to as digital data systems. Many types of data systems, of varying complexity, are used in the biomedical field as well as all other fields. The major functional elements of digital data systems are illustrated in the example shown in Figure 1-3. A *multiplexer* is used to sequentially sample the conditioned outputs of sensors **1** through **n.** The earliest form of a multiplexer was the electromechanical commutator, essentially a motor-driven rotary selector switch whose arm would dwell sequentially on contacts to which the various sensor outputs were connected. The switch cycles through the contacts rapidly enough to get a sufficient number of data points for each sensor signal. Relatively rapidly varying sensor outputs are sampled more often than relatively slowly varying outputs. The dwell time on each contact, the time to switch between contacts, and the sequence of contacts are all known and this allows reconstruction of sensor signals from the sampled signals appearing sequentially on a common line. Modern electronic logic circuitry has obsoleted the electromechanical commutator and permits much greater flexibility in sampling and sequencing, but the basic principle remains the same.

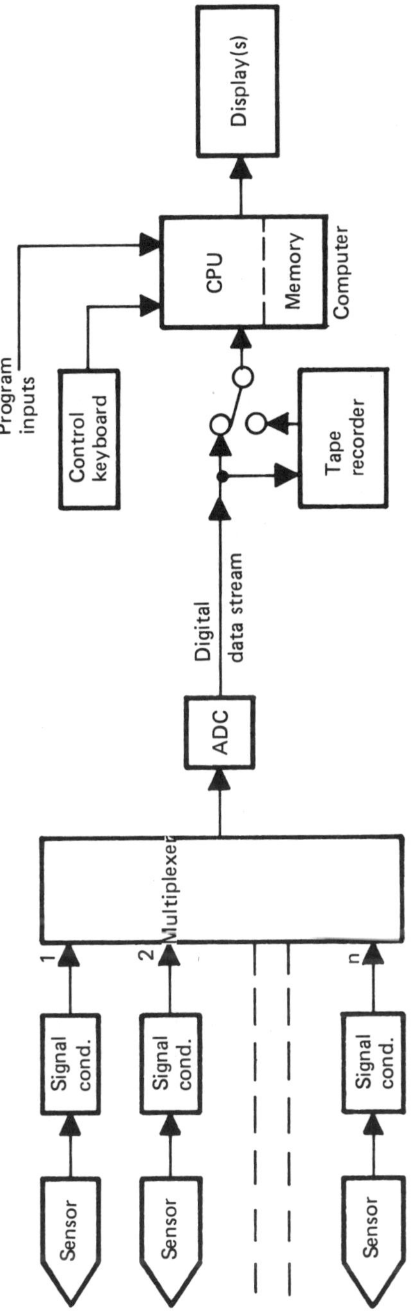

Figure 1-3. Example of multisensor digital data system

Many data systems use a single analog-to-digital converted (ADC) to convert the analog samples on the common line into digital form. All sensor signals must then be conditioned so that the output range is compatible with the input range of the ADC. The output of the ADC is a digital data stream. When the digital data words (corresponding to sensor-output data points) must be arranged in a specific format, or when additional information must be included in this data stream (e.g., station number, time of day) a *formatter* is used for these purposes.

The digital data stream from the ADC (or from the formatter added to the ADC) is then fed to the "computer" portion of the data system. Most systems include a digital tape recorder so that the incoming data need not be processed immediately (*real-time* processing) but can be stored and then played back and processed later (*non-real-time* processing). The two most essential elements of the computer are the central processing unit (CPU) and the memory (core). The computer can be programmed to perform a variety of data processing functions and, with the aid of a control keyboard, specific forms of processing can be selected, additional information can be entered, and various forms of displays, on different types of display devices, can be obtained.

The most frequent use of biomedical sensors is in measuring systems (which include *monitoring* systems). However, some applications in automatic control systems do exist, and the fundamentals of such systems are briefly explained below.

1.3 CONTROL SYSTEMS

There are two basic types of control systems: *open-loop* control systems include a human operator who observes information displays originating from sensors and then effects a control function manually, e.g., increases or decreases gas flow, increases or decreases a pressure, or adjusts a flow rate; *closed-loop* control systems provide *automatic control* over a quantity without the actions of a human operator (although operators are often employed to monitor the proper functioning of such systems); instead, the conditioned output of a sensor is used to effect a control function directly, by means of other electronic and electromechanical devices.

A simple closed-loop control system is illustrated in Figure 1-4. The quantity to be controlled is acted upon by a control device so that the quantity is then controlled at some predetermined value. A sensor provides a signal indicative of the value of the controlled quantity. The conditioned sensor signal is fed to a regulating device in which it is compared against a reference signal. The reference signal has been adjusted, e.g., by manual rotation of a potentiometer having a calibrated dial, so that it equals that value of (conditioned) sensor output that would correspond to the desired value at which the quantity is intended to remain. If the sensor signal equals the reference signal, the quantity is controlled at its proper value and no action is required from the control device. If, however, the sensor signal is higher or lower than the reference signal, the regulating device acts upon the difference-signal (and its polarity or phase) by sending a signal to the control device to increase or decrease, as appropriate, the value of the quantity until it is again at the desired value. When this point has been reached, the sensor signal will again equal the reference signal and control-device action will stop.

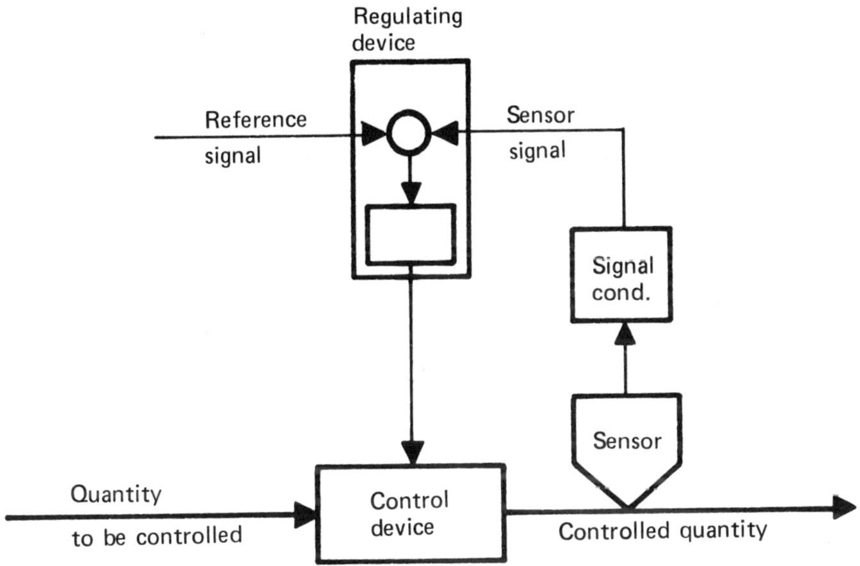

Figure 1-4. Basic control system

The most common form of closed-loop control is *proportional control* in which the regulator action is proportional to sensor/reference signal difference; hence, a small signal difference will cause a small regulating action, and a large signal difference will cause a large regulating action.

Typical examples of closed-loop control are: flow control, in which the sensor is a flowmeter, the quantity to be controlled is liquid or gas flow, and the control device is an electrically-driven control valve; temperature control, in which the sensor is a thermocouple or resistive temperature sensor, the quantity to be controlled is the temperature (e.g., of an autoclave or temperature bath) and the control device is an electric heating element; and pressure control, in which the sensor is a pressure transducer, the quantity to be controlled is the pressure of a liquid or gas, and the control device is an electrically-driven pressure regulator.

1.4 MEASURANDS AND UNITS

Sensors have been developed for measurements of numerous different quantities, and the *measurand* (for quantity intended to be measured by the sensor) is one of the essential descriptors of a sensor. Thus we speak of pressure transducers, humidity sensors, flowmeters (flow sensors), load cells (force sensors) and gamma-ray detectors (gamma-ray sensors). The measurands can be grouped into a relatively few major categories. An essential adjunct to any given measurand is the unit in which it is expressed. This section defines the measurands that are commonly measured by sensors, explains some typical sensing techniques, and shows the units of measurement in which the measurands are expressed.

1.4.1 Solid-Mechanical Quantities

These include: *position* of an object or point with respect to a reference point; *displacement* or *motion,* which are changes in position and can be either *angular* (e.g., a 110 degree displacement of the forearm) or *linear* (e.g., a 5 cm upward motion, or displacement, of the shoulders); *velocity,* the time rate of change of displacement (motion), which can also be angular (e.g., a velocity of 6 degrees per second) or linear (e.g., a velocity of 80 kilometers per hour); *acceleration,* the time rate of change of velocity, which can also be angular (e.g., an acceleration of 0.4 degree per second per second) or linear (e.g., an acceleration of 9.8 meters per second per second); *attitude,* a relative angular orientation of an object with respect to a set of orthogonal reference axes (e.g., an inclination of 5 degrees from gravity-based vertical, toward the right); *strain,* the deformation of a solid resulting from mechanical stress (e.g., a strain of 0.5 millimeter per meter); *mass*, the inertial property of a body, i.e., a measure of the quantity of matter in a body and of its resistance to changes in motion; mass measurement is commonly called "weighing" (e.g., a mass of 85 kilograms); on Earth, "weight" is the gravitational force with which a body is attracted toward the Earth; *force*, the vector quantity needed to cause a change in momentum; it is the product of mass and acceleration (F = m·a); it is properly expressed in newtons (e.g., a force of 1.5 newtons) but was formerly expressed in pounds (force) or kilograms (force); and *torque,* the moment of force, and the product of force and the length of the lever arm (e.g., when a force of 1 newton is applied at the end of a lever arm that has its other end attached at the axis of rotation and is 1 meter long, the torque is 1 newton-meter).

It should be noted that *vibration* and *shock* (mechanical shock) are generally measured with acceleration sensors (called *accelerometers*), that strain sensors are generally called *strain gages,* that force sensors are often referred to as *load cells,* and that angular velocity sensors, when used to measure rotational speed (e.g., of a shaft or of an agitator paddle) are called *tachometers.*

1.4.2 Fluid-Mechanical Quantities

Fluid-mechanical quantities are: *flow,* the motion of a fluid (which can be liquid or gaseous), which is generally sensed as *flow rate,* the time rate of motion of a fluid (e.g., a flow rate of 500 milliliters per hour, or of 200 grams per hour; the former is a *volumetric* flow rate, the latter a *mass* flow rate); *density,* the ratio of mass to volume of a substance (e.g., a density of 0.4 kilogram per cubic meter); *viscosity,* the resistance (of a fluid) to the tendency to flow (e.g., a viscosity of 0.04 newton-second per square meter); *humidity,* the amount of water vapor in a gas; usually it is *relative humidity* that is sensed (the ratio of water-vapor pressure to that amount of water-vapor pressure that would be required for saturation, e.g., a relative humidity of 85%); *moisture,* the amount of absorbed or adsorbed water on a solid or in another liquid (expressed in percent by weight or percent by volume); the *dew point,* the temperature at which the relative humidity is 100%, i.e., at which water-vapor saturation occurs (e.g., a dew point of 18°C); any cooling below this temperature would produce water condensation (a dew); *liquid level,* the height of the surface of a liquid or quasi-liquid (e.g., powdered or granular solids) above the bottom of the container in which the liquid is kept (or above another reference point if so specified); and *pressure,* the force of a fluid acting on a unit area (e.g., a pressure of 550 kilopascals; the *pascal* is the name given to the unit *newton per square meter*).

Pressure can be measured relative to zero pressure (a perfect vacuum) and is then called *absolute* pressure; it can be measured with reference to the ambient pressure and is then called *gage* pressure; or it can be the pressure difference between two points of measurement (*dif-*

ferential pressure). The term *vacuum* generally applies to pressures below one kilopascal (below about one-hundredth of normal atmospheric pressure), and vacuum sensors are generally quite different in design than pressure sensors.

It should be noted that flow-rate sensors are usually called *flowmeters,* density sensors are often referred to as *densitometers* (not to be confused with optical densitometers that are used, e.g., to determine the relative opacity of portions of exposed photographic or X-ray film), viscosity sensors are sometimes called *viscometers,* relative-humidity sensors are known as *hygrometers* or *psychrometers* depending on type of sensor used, pressure sensors are more often referred to as *pressure transducers,* and vacuum sensors are frequently called *vacuum gages.*

Pressure can also be considered to include the acoustic quantities *sound pressure,* the instantaneous pressure, at a given point, due only to the presence of a sound wave; *sound pressure level,* defined as 20 times the logarithm (to the base 10) of the ratio of an rms sound pressure to an rms (specified) reference pressure, and expressed in decibels; and *sound level,* a sound pressure level that is weighted in a manner, and by means, specified by a national or international standard.

1.4.3 Thermal Quantities

The main quantity of interest here is *temperature,* the thermal state of a body considered with reference to its power of communicating heat to other bodies. Temperature sensors are commonly referred to as *thermometers,* particularly when they are placed in situ, i.e., held in contact with the object whose temperature is to be measured. There also exist noncontacting thermometers (remote-sensing temperature sensors) which are pointed at the object to be measured from a distance and are also known as *radiation pyrometers.*

Another thermal quantity for which sensors are available is *heat flux,* the time rate of transfer of heat energy (expressed in watts). Heat can be transferred by one or more of the following: conduction, convection and radiation. *Calorimeters* sense the combined radiant and convective heat flux; *radiometers* sense only radiant heat flux, and some types of *surface heat-flow sensors* can sense a combination of all three types of heat transfer.

1.4.4 Electromagnetic and Particle Radiation

These quantities include *visible light, ultraviolet* (UV) and *infrared* (IR) *light, gamma-rays, X-rays, alpha and beta particles, protons, and neutrons.* The electromagnetic spectrum extends upward in wavelength (downward, in frequency) to submillimeter and millimeter (mm) waves, microwaves, radio waves, and sonic (sound) and infrasonic (below the audible) waves; however, sensors for those quantities are quite different in nature from those used to sense particle radiation and electromagnetic radiation at wavelengths below about 10^{-4} m. The electromagnetic spectrum (for wavelengths of 1 m and less) is illustrated in Figure 1-5. It could be noted that the spectrum of ultrasonic waves extends approximately from 10^2 to 5×10^4 m (not included in the illustration).

Radiation from far-IR to far-UV is sensed in terms of *intensity* and *wavelength* (or intensity over a band of wavelengths). Radiation below the far-UV portion of the spectrum is sensed in terms of intensity and *energy* (or intensity over a band of energy levels). Radiation energy (photon energy) scales with frequency (hence, inversely with wavelength) and ranges between 10 electron volts at a wavelength of about 100 nanometers (extreme UV) to 10^9 electron volts at a wavelength of about 10^{-6} nanometers (cosmic rays); this is the reason for calling X-rays of relatively longer wavelengths (lower energy) "soft" and calling X-rays of relatively shorter wavelengths (higher energy) "hard." Nuclear (particle) radiation is sensed in terms of number of particles (or number of particles per unit time), called *count,* and energy (or count over a band of energy levels).

Sensors for these quantities are more commonly called *detectors.* Electromagnetic-radiation detectors are generally classified as either *photon detectors* or *thermal detectors.* Photon detectors respond to incident radiation quanta (photons) that react with electrons in the sensor material. Thermal detectors, used primarily for IR sensing, respond to the total incident radiant energy. Nuclear radiation detectors, in general, employ *ionization* (in a gas, between two electrodes, or in a semiconductor material) or *scintillation* (in a material that then produces photons that can, in turn, be detected by a photon detector) for their operation.

1.4.5 Electric and Magnetic Quantities

These quantities include *current, voltage, power, electrostatic charge,* and *magnetic flux density.* Many of the sensor types used in this field employ the Hall effect for their operation.

1.4.6 Chemical Properties

Only a few types of instruments used for determinations of chemical properties and composition can be called sensors. The majority are analyzers, or analysis instruments, used for *qualitative analysis* ("What elements are contained in this sample?") and *quantitative analysis*

("How much of one or more given elements are contained in this sample?"). Those devices that are used in some areas of electroanalytic chemistry are called sensors: *conductivity* sensors measure solution concentrations or the amounts of a salt in aqueous solution; *pH sensors* measure the acidity or alkalinity of a solution; *ORP sensors* measure the oxidation or reduction potential (ORP, or *redox*) of a solution; *specific-ion sensors* respond to the relative abundance of ions of selected substances (e.g., metals, ammonia, nitrate, sulfate). Other analyzers which are based on the electrical characteristics of an *electrochemical cell* are the *coulometer,* the *polarograph,* and electrometric gas analyzers.

Thin-film *resistive gas sensors,* using metal-oxide films, have been developed as oxygen analyzers as well as for measuring concentrations of other gases. Various types of *ionization analyzers* are also used for gas analysis, as are *thermal conductivity cells.* Both of these types of analyzers are used as detectors in *gas chromatographs* which are used for qualitative and quantitative analyses of gaseous or liquid mixtures.

Spectroradiometric analyzers, which include *spectrometers* and *spectrophotometers,* are widely used for chemical composition analyses. Such instruments operate in almost all portions of the electromagnetic spectrum shown in Figure 1-5, though each type of spectrometer is designed for only one region of the spectrum: spectrophotometers measure spectra (electromagnetic radiation intensity vs wavelength) in the visible and adjacent UV and IR (near-UV to near-IR) bands; submillimeter and microwave spectrometers measure in their wavelength bands; IR spectrometers operate in the near and far infrared bands; gamma-ray and X-ray spectrometers measure spectra (intensity vs energy) in their respective wavelength/energy bands. Chemical composition is determined from analysis of the spectral lines (line height is given by intensity, line position by wavelength) of the *spectrograms* obtained from these instruments, due to either radiation emission or absorption by substances.

Different principles are used in other spectrometers that are also used for compositional analyses. Magnetic resonance spectra are provided by nuclear magnetic resonance (NMR) and electron spin resonance (ESR) spectrometers. The spectra produced by a *mass spectrometer* allow compositional analysis from a display of the mass/charge ratio of ions. Spectrometers used in the relatively new field of *surface analysis* respond to the energies of electrons and ions from samples that are bombarded with a narrow beam of electrons, X-rays, ultraviolet rays, or ions.

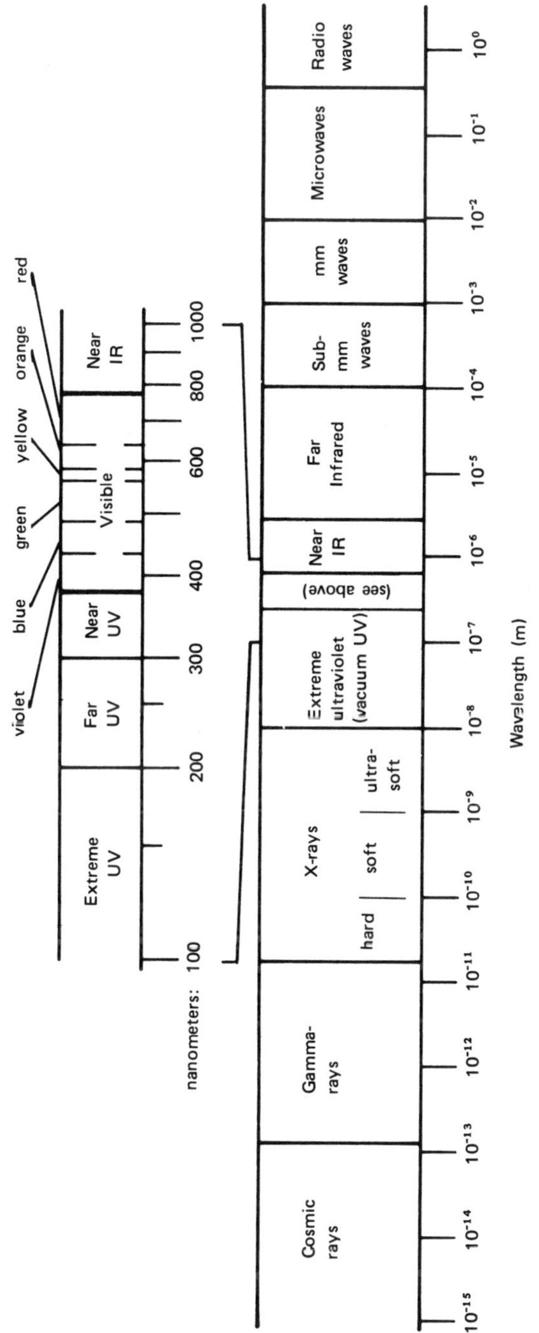

Figure 1-5. The electromagnetic spectrum (for wavelengths between 1 and 10^{-15} m)

1.4.7 Units of Measurement

The units in which the many different measurands are expressed, have had various names, magnitudes, and definitions, depending on which system of units is in use. The U.S. Customary System (formerly called "English system") of units is still in use in the U.S.A., although an Act of Congress made the use of "weights and measures of the metric system" legal in 1866.

One or the other form of a metric system has been in use in most other countries of the world. For many countries that meant a change from at least portions of an older system.

The International System of Units (SI, from the French *Système International d'Unités*), a rationalized and coherent system of units based on metric units, is now the recommended system of units for all countries. It has been legally adopted by most countries and can be expected to be used world-wide in the fairly near future. An effort has been made, in this book, to use SI units to the greatest extent possible.

Table 1-1 is a compilation of SI units; it includes conversion factors for some non-SI units; it shows the symbols for these units; it also includes some commonly-used non-SI units and their symbols. Table 1-2 shows the prefixes (and their symbols) used for multiple and submultiple units.

An explanation is necessary for the spelling of "meter" and "liter." The preferred spelling is "metre" and "litre;" this spelling is based on the French spelling and was also adopted in the U.K.; it is strongly supported by many associations and societies, including the U.S. Metric Association (USMA) of which the author is a member. However, the U.S. Department of Commerce prescribed the "German" spellings "meter" and "liter" for use in all work by and for the U.S. Government and the author subjectively prefers this spelling. In general, unless constrained by an applicable regulation prescribing one or the other form of spelling, the use of either spelling should be considered satisfactory.

Many SI units have been in use in medicine and biology for a long time, including in the U.S.A., and others will hopefully be adopted soon. Some of the symbols for such units, or submultiples of units, are not always used in their correct form; the worst offender is probably the "cc" that is used instead of correct cm^3." It is hoped that the correct use of symbols for units of measurement will also be implemented soon.

Table 1-1: Units, Symbols, and Conversion Factors

Quantity	SI Unit	SI Symbol	Non-SI Unit	Non-SI Symbol	To Convert from the Non-SI to SI Unit, Multiply it by:
Solid Mechanical Quantities					
Acceleration, linear	meter/second^2	m/s^2	(free fall, std.)	g	9.8
Acceleration, angular	radian/second^2	rad/s^2			
Angle, plane	radian	rad	degree	°	1.745×10^{-2}
			minute	'	2.909×10^{-4}
			second	"	4.848×10^{-6}
Angle, solid	steradian	sr			
Area	meter^2	m^2	foot^2	ft^2	9.29×10^{-2}
			mile^2	mi^2	2.59×10^{6}
			acre		4.047×10^{3}
Force	newton	N	pound-force	lb_f	4.448
			kilogram-force	kg_f	9.80665
Length	meter	m	inch	in	2.54×10^{-2}
			foot	ft	0.3048
			yard	yd	0.9144
			mile (statute)	mi	1.609×10^{3}
Mass	kilogram	kg	ounce (mass)	oz	2.835×10^{-2}
			pound-mass	lb_m	0.4536
			grain	gr	6.48×10^{-5}
Strain			microstrain	$\mu\epsilon$	

(continued)

Table 1-1: (continued)

	SI		Non-SI		To Convert from the Non-SI to SI Unit,
Quantity	**Unit**	**Symbol**	**Unit**	**Symbol**	**Multiply it by:**
Torque	newton-meter	N·m	ounce-inch	oz_f-in	7.06×10^{-3}
			dyne-centimeter	dyn-cm	1.00×10^{-7}
Velocity, linear	meter/second	m/s	inch/second	in/s	2.54×10^{-2}
			mile/hour	mi/h	0.447
Velocity, angular	radian/second	rad/s	degree/second	°/s	1.745×10^{-2}
			revolution/ minute	r/min	0.1047
Volume	meter^3	m^3	foot^3	ft^3	2.832×10^{-2}
	liter (= dm^3)	ℓ or L	fluid ounce	fl oz	2.957×10^{-2}
			quart	qt	0.946
			pint	pt	0.473
			gallon	gal	3.7854
	(centimeter^3	cm^3)	(fluid ounce	fl oz	29.57)
Fluid-Mechanical Quantities					
Density	kilogram/meter^3	kg/m^3	ounce/gallon	oz/gal	7.489
			slug/foot^3	slug/ft^3	515.38
Flow(-rate) (mass)	kilogram/second	kg/s	pound/minute	lb_m/min	7.56×10^{-3}
Flow(-rate) (volumetric)	meter^3/second	m^3/s	gallon/minute	gal/min	6.309×10^{-5}
			foot^3/minute	ft^3/min	4.7195×10^{-4}
	liter/second	ℓ/s, L/s	gallon/minute	gal/min	6.309×10^{-2}

(continued)

Table 1-1: (continued)

Quantity	SI Unit	SI Symbol	Non-SI Unit	Non-SI Symbol	To Convert from the Non-SI to SI Unit, Multiply it by:
Pressure	pascal (= N/m^2)	Pa	atmosphere (std)	atm	1.01325×10^5
			bar	bar	10^5
			inch of mercury	in Hg	3.3864×10^3
			inch of water	in H_2O	2.491×10^2
			kilogram/cm^2	kg_f/cm^2	9.80665×10^4
			pound/$inch^2$	lb_f/in^2	6.895×10^3
			torr	torr	133.32
Viscosity, dynamic	newton-second/ $meter^2$	$N \cdot s/m^2$	centipoise	cP	10^{-3}
Viscosity, kinematic	$meter^2$/second	m^2/s	centistokes	cSt	10^{-6}
Thermal Quantities					
Temperature	kelvin	K	degree Celsius	°C	K = °C + 273.15
			degree Fahrenheit	°F	K = (°F + 459.67)/1.8
	degree Celsius	°C	degree Fahrenheit	°F	°C = (°F - 32)/1.8
Heat energy	joule	J	British thermal unit	Btu	1.055×10^3
Heat power	watt	W	Btu/hour	Btu/h	0.293
Heat flux	watt/$meter^2$	W/m^2	calorie/cm^2-second	$cal/cm^2 \cdot s$	4.184×10^4
Radiant intensity	watt/steradian	W/sr			

(continued)

Table 1-1: (continued)

Quantity	SI Unit	SI Symbol	Non-SI Unit	Non-SI Symbol	To Convert from the Non-SI to SI Unit, Multiply it by:
Light					
Luminous intensity	candela	cd	candlepower	cp	1.00
Luminous flux	lumen	lm			
Illumination	lux	lx	footcandle	fc	10.764
Luminance	candela/meter^2	cd/m^2	footlambert	fL	3.4263
Nuclear Radiation					
Energy	joule	J	electron volt	eV	1.6021×10^{-19}
Radioactivity	bequerel (= s^{-1})	Bq	curie	Ci	3.70×10^{10}
Absorbed dose	gray	Gy	rad	rad	10^{-2}
Exposure dose	coulomb/kilogram	C/kg	roentgen	R	2.58×10^{-4}
Dose equivalent	sievert	Sv	rem	rem	10^{-2}
Electric and Magnetic Quantities					
Capacitance	farad	F			
Field strength	volt/meter	V/m			
Inductance	henry	H			
Resistance	ohm	Ω			

(continued)

Table 1-1: (continued)

Quantity	SI Unit	SI Symbol	Non-SI Unit	Non-SI Symbol	To Convert from the Non-SI to SI Unit, Multiply it by:
Frequency	hertz	Hz	cycle/second	c/s	1.00
Power	watt	W			
Quantity of electricity	coulomb	C	ampere-hour	A·h	3.60×10^{3}
Current	ampere	A			
Voltage, emf	volt	V			
Conductance	siemens	S	reciprocal ohm	mho	1.00
Magnetic flux	weber	Wb	maxwell	Mx	1.00×10^{-8}
Magnetic flux density	tesla	T	gauss	G	1.00×10^{-4}
Magnetic field strength	ampere/meter	A/m	Oersted	Oe	79.58
Other Quantities.					
Substance (chem.)	mol	mol			
Time	second	s	minute	min	60
			hour	h	3600
			day	d	8.64×10^{4}
Digital data-word length			bit	b	
Digital data rate			bit/second	b/s	

Table 1-2: Multiple and Submultiple Prefixes for Units

Multiplication Factor	Prefix	SI Symbol	Multiplication Factor	Prefix	SI Symbol
10^{18}	exa-	E	10^{-1}	deci-*	d
10^{15}	peta-	P	10^{-2}	centi-*	c
10^{12}	tera-	T	10^{-3}	milli-	m
10^{9}	giga-	G	10^{-6}	micro-	μ
10^{6}	mega-	M	10^{-9}	nano-	n
10^{3}	kilo-	k	10^{-12}	pico-	p
10^{2}	hecto-*	h	10^{-15}	femto-	f
10^{1}	deka-*	da	10^{-18}	atto-	a

*Nonpreferred and not recommended; use now restricted to: hectare (=10^4 m^2), hectoliter, dekagram, deciliter, decibel (dB), and centimeter.

1.5 SENSING ELEMENTS

A sensor (or transducer) is, by definition, a device which provides a usable output in response to a specific measurand. Hence, it must provide two essential functions (see Figure 1-6): it must *sense* the measurand, and it must produce an output (electrical output) that is indicative of the measurand sensed, i.e., it must provide for the *transduction* of the measurand.

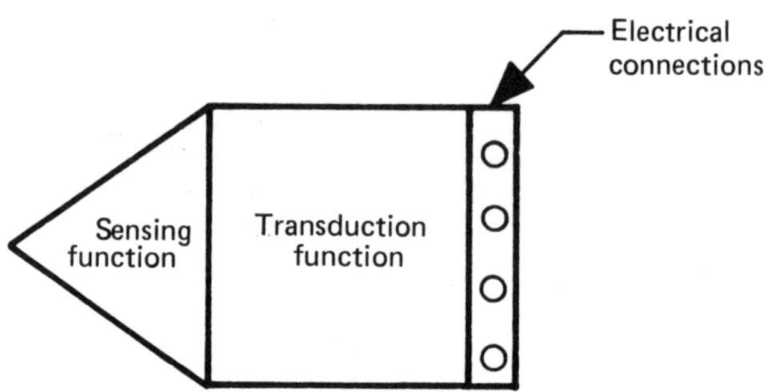

Figure 1-6. Generalized schematic representation of a sensor

In some types of sensors the sensing function and the transduction function are performed by separate elements (see Figure 1-7a). One element is that portion of the sensor that responds directly to the measurand *(sensing element)*. The other element is that portion of the sensor in which the output originates *(transduction element)*; the nature of the operation of the latter is called the *transduction principle* of the sensor.

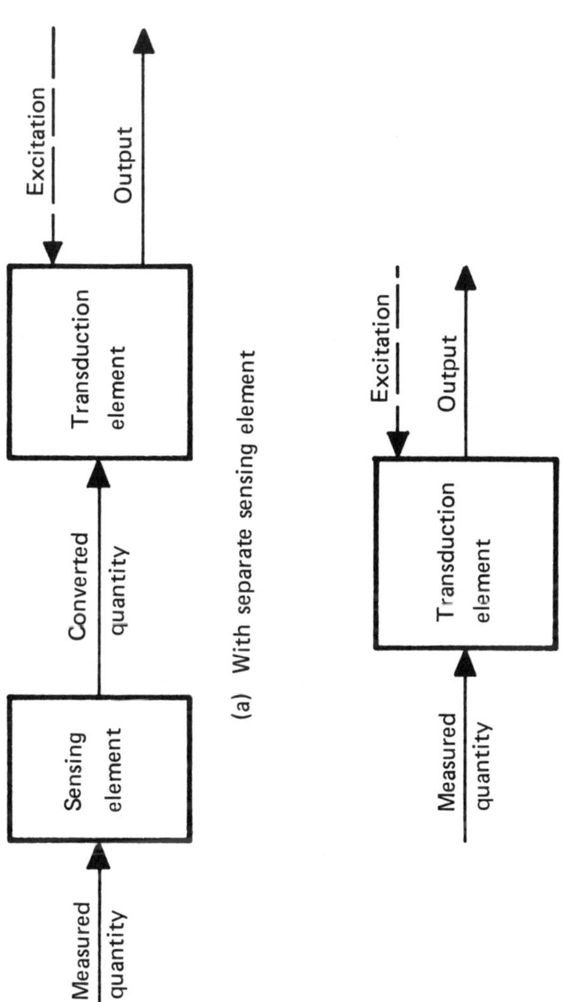

(a) With separate sensing element

(b) With sensing and transduction provided by the same element

Figure 1-7. Sensor basic functional flow diagram

Transduction principles are covered in Section 1.6. In other types of sensors the sensing and transduction functions are both provided by the same element (see Figure 1-7b). Examples of such dual-function elements are the thermocouple and the resistance elements (wire-wound or thermistor type) that respond directly to temperature, when temperature is the measurand. Both of these generic classes of sensors (those with separate and those with dual-function elements) may or may not require external electrical power *(excitation)* for their operation, depending on whether the transduction element is of the self-generating or non-self-generating type (see Section 1.6).

Sensing elements are used to facilitate the transduction of certain measurands. Examples are illustrated in Figure 1-8. Pressure sensors employ a sensing element, such as a metal diaphragm, to convert pressure into strain or a displacement (due to the mechanical deflection of the diaphragm under pressure); the transduction element then responds to either the displacement or the strain. Some types of flowmeters employ a turbine-type sensing element. The rotational speed of the turbine is proportional to flow and the transduction element responds to the rotational (angular) velocity of the turbine element. Scintillation counters, which are often used for nuclear radiation detection, employ a scintillator material as sensing element. The scintillator (which can be liquid or solid) responds to incoming nuclear radiation by producing photons (light); the light is then transduced by a photoelectric (photoconductive, photovoltaic, photoemissive, etc.) transduction element. Other examples are the seismic mass in an accelerometer (a small weight that responds to acceleration by producing displacement or strain) and the deflecting-beam, proving-ring, or column-type sensing element used in force sensors (load cells) that produces strain (sometimes displacement) when a force acts on it.

1.6 TRANSDUCTION PRINCIPLES

The transduction principle of a sensor is the operating principle of its transduction element; as explained in 1.5, the transduction element also provides the sensing function in many types of sensors. The most commonly used transduction principles are described here, together with some remarks concerning the nature and operation of the corresponding transduction elements.

1.6.1 Self-Generating Elements

This category of transduction elements does not require external electrical excitation. Their operation is based on certain physical effects

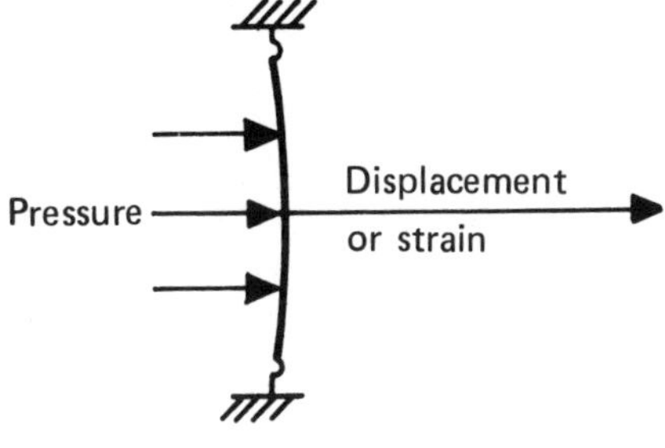

(a) Diaphragm

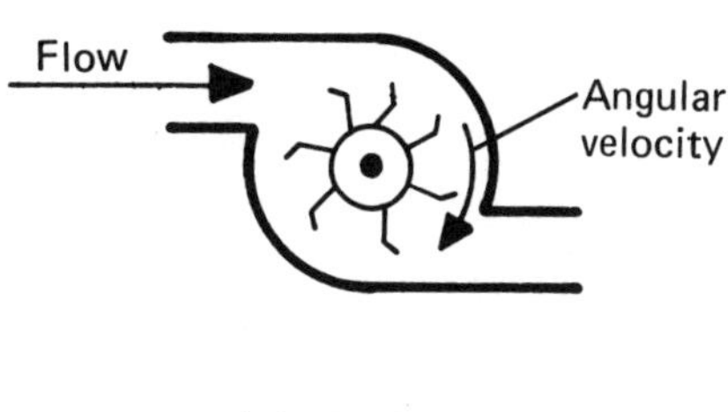

(b) Turbine

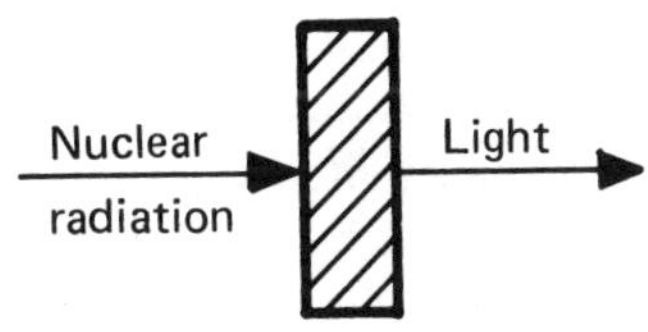

(c) Scintillator

Figure 1-8. Examples of sensing elements

that were discovered to occur in certain materials under certain conditions.

1.6.1.1 Thermoelectric Transduction: The thermoelectric effect, known as the *Seebeck effect,* is illustrated by the *thermocouple circuit* shown in Figure 1-9a. Two conductors, made of dissimilar metals, are joined together at their *sensing junction* (J_s) and through a load resistor (R_L) at their reference junction (J_r). When the sensing junction temperature (T_1) is higher than the reference junction temperature (T_2), a current (I) will flow through the circuit in the direction indicated by the arrow. If a voltmeter were connected across R_L, it would indicate a voltage that would increase as T_1 increases, provided that T_2 does not change. In this circuit, Metal A is selected so that it produces a positive thermoelectric potential (vs a reference metal, e.g., platinum), and Metal B is selected so that it produces a negative potential. If the same circuit were used under conditions where T_1 is lower than T_2, the current flow would be in the opposite direction.

Thermocouples are mostly used for temperature measurement, although the thermoelectric effect is also used in some sensors for other measurands. They typically consist of two wires welded together at their sensing junction. The sensing junction is held firmly to the object whose temperature is to be measured. The other ends of the two wires are kept at a known and, preferably, controlled temperature. The thermocouple wires are made of some pure metals and some special metal alloys. One type of thermocouple, e.g., has one wire made of platinum, the other of a platinum/rhodium alloy. The output voltage produced by thermocouples typically ranges between millivolts and tens of millivolts. Standard tables are available that correlate output voltage with temperature for various metal combinations.

Multiplication of output voltage can be obtained by connecting several thermocouples in series, at their colocated reference junctions; this arrangement is called a *thermopile.* It should be noted that the reverse thermoelectric effect, called *Peltier effect,* is obtained when a source of electrical power is connected to the two terminals at the reference junction. Depending on the polarity of the power, the other junction will then be heated above, or cooled below, the temperature at the reference junction. This principle is used in thermoelectric coolers.

1.6.1.2 Piezoelectric Transduction: Certain materials, e.g., quartz crystal, exhibit the piezoelectric effect: they produce electricity across their surfaces when mechanically stressed (see Figure 1-9b). The mechanical stresses can be induced by compression or bending of the crystal. Quartz is the only natural crystal material that is still widely used. Many piezoelectric elements (ceramic crystals) are made of ce-

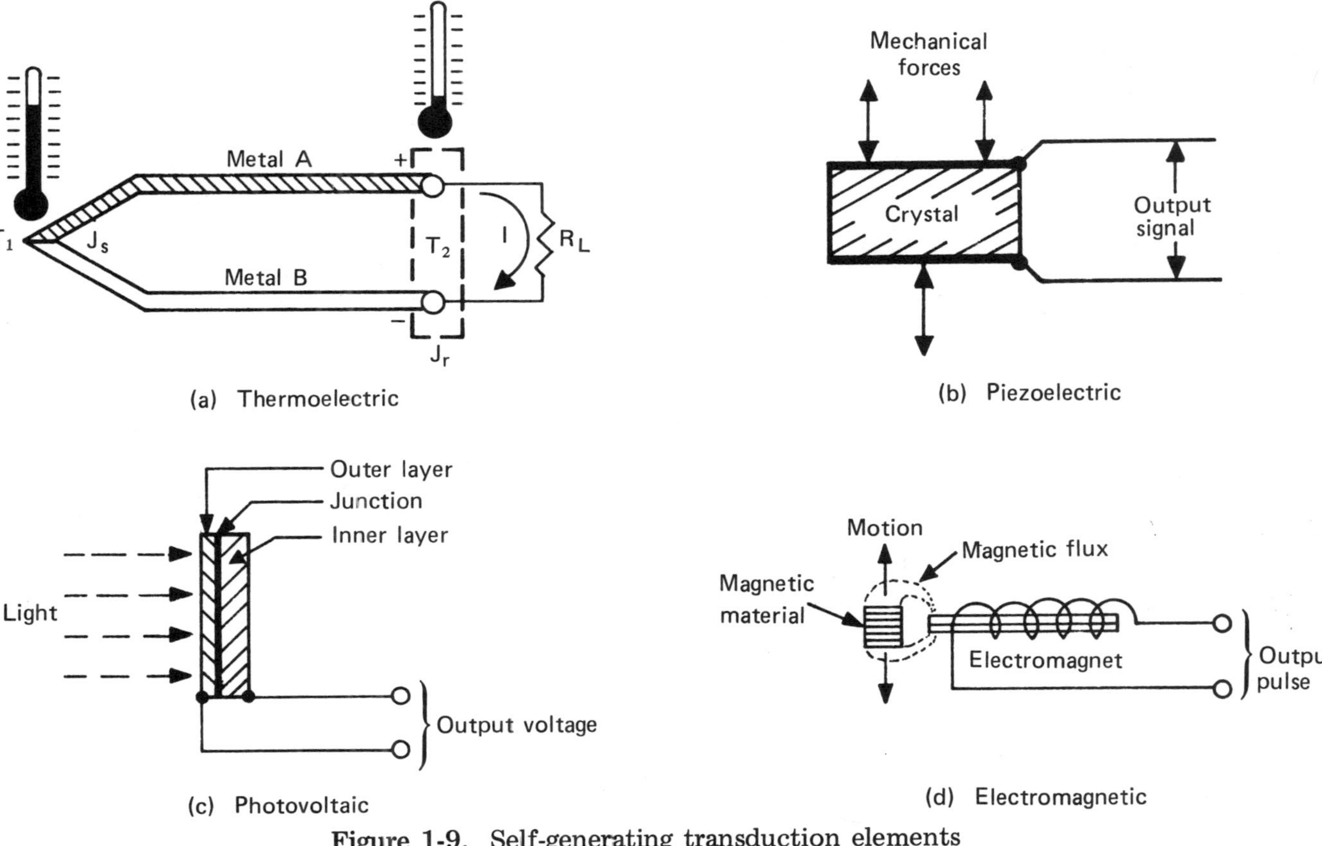

Figure 1-9. Self-generating transduction elements

ramic which is artificially polarized to produce this effect. The output impedance of such crystals is very high; hence, an impedance converter or amplifier is usually coupled to the crystal with a short cable or, in miniature solid-state form, mounted in the same housing as the crystal.

Piezoelectric elements can be used only when the measurand fluctuates (at rates above about 1.5 Hz), not when steady-state measurand levels are to be transduced. They are widely used in microphones, pressure transducers (when the pressure fluctuates), accelerometers (especially for vibration measurements) and force sensors (for fluctuating forces), as well as in ultrasonic probes. The reverse piezoelectric effect (mechanical deformation in response to an applied ac voltage) can also be obtained in such crystals. Hence, they can be used for the sending as well as receiving of sound, including ultrasound.

1.6.1.3 Photovoltaic Transduction: A basic photovoltaic element (photovoltaic cell) consists of two different semiconductor materials, one forming a thin and essentially transparent outer layer, the other forming an inner layer. When the junction between these two layers is illuminated, a voltage is produced across the "sandwich" made up of the two layers. This voltage increases with increasing illumination. A popular example of such an element is the (photovoltaic) solar cell that is used to convert sunlight into electrical power. Figure 1-9c illustrates a photovoltaic transduction element.

1.6.1.4 Electromagnetic Transduction: An electromagnet consists of a ferromagnetic (e.g., iron) core around which many turns of wire are wound. When a voltage is applied across the coil, a magnetic field is created at the ends of the core. When used in a transduction element, however, the electromagnet is used in a passive mode and the magnetic field is set up by a permanent magnet that is either used as the core or is bonded to a (ferromagnetic) core. When the magnetic field is disturbed by a piece of magnetic material moving through it, a voltage is induced in the coil. If the motion of the piece is in one direction the voltage will first rise, then decay, i.e., it will be a voltage pulse. Figure 1-9d illustrates electromagnetic transduction.

A typical use of such an element is in a tachometer which has a "tooth" of magnetic material attached to the rotating shaft. The electromagnet ("pickoff coil") is mounted so that this tooth passes it, and disturbs its magnetic field once every revolution. By counting the number of pulses produced per minute, the number of shaft revolutions per minute can be established. The same principle is used in some types of turbine flowmeters in which the turbine blades are either made of a magnetic material or have small magnets imbedded in their tips. As the blade tips pass by an electromagnet, pulses are generated, and the number of pulses per unit time are proportional to the flow rate.

1.6.2 Non-Self-Generating Elements

Passive electrical elements such as variable capacitors, inductors or resistors comprise or are used in most of this category of transduction elements, which all require some form of externally-supplied electrical excitation for the operation.

1.6.2.1 Resistive Transduction: Conductors as well as semiconductors have *resistance* (resistance to current flow through them). There are a number of effects that occur in certain materials that cause this resistance to change in response to measurand changes (see Figure 1-10a).

Temperature–Conductors (metal wire, metal film foil) change their resistance when they are heated or cooled. This effect is widely used in *resistance thermometers;* they employ, as their sensing/transduction elements, a winding of fine wire (typically platinum) or a (platinum) film, sometimes a thin metal foil. Conductors generally increase in resistance with increasing temperature. Certain semiconductor materials also change their resistance when heated or cooled. The best known of these elements is the *thermistor;* it changes its resistance inversely with temperature and this *R-vs-T* relationship is usually nonlinear; however, it is reasonably linear in the range of temperatures measured by medical thermometers. The R-vs-T relationship is much more linear for conductors; however, the resistance change ($\Delta R/R/^{\circ}C$) is significantly less in conductors than in thermistors.

Humidity–Certain materials change their resistance when the humidity they are exposed to undergoes changes. One example of such *hygrometric* elements contains a hygroscopic film of *lithium chloride* which responds to humidity changes with resistance changes. Another example is an element made of *sulfonated polystyrene* which changes its surface resistance with humidity.

Strain–Conductors as well as certain semiconductors change their resistance when they are subjected to mechanical deformation (tension or compression). This effect is known as *piezoresistance* (the Greek word *piezein* means "squeeze"); this term has been (incorrectly) applied to the effect when it occurs in semiconductors only. The effect is used primarily in *strain gages.* Metal-wire or metal-film gages typically consist of a thin *grid* (actually more of zig-zag pattern) of a metal that is relatively free of resistance changes due to temperature (otherwise their response to temperature changes may obscure their response to strain changes).

Semiconductor strain gages are made strain-sensitive by appropri-

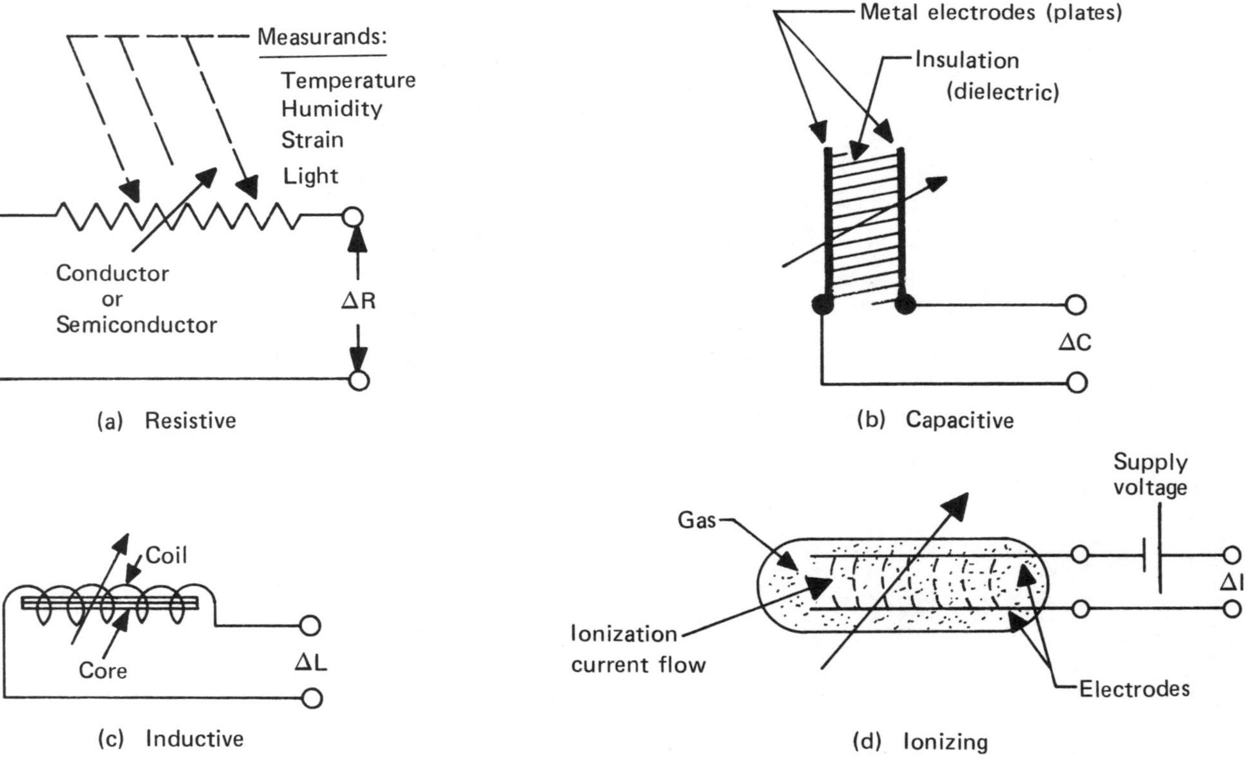

Figure 1-10. Basic passive transduction elements

ate doping of the semiconductor (usually silicon) material. The resistance changes of semiconductor gages are more pronounced than those of metal foil or wire gages; however, it is easier to keep temperature sensitivity very low in metal gages than in semiconductor gages. Strain gages (sometimes first bonded to a thin insulating substrate) are usually cemented to the surface in which strain is to be measured, following carefully-controlled procedures.

Light–Certain semiconductor materials change their resistance when the incident illumination changes *(photoconductive elements)*. The most widely used photoconductors are cadmium sulfide and cadmium selenide; they are very sensitive in the visible-light region and are relatively easy to produce. Lead sulfide and lead selenide elements are more suitable for near-infrared measurements. Other photoconductors such as gold-doped germanium and mercury-doped germanium can be used for far-infrared sensing (to about 14 μm for mercury-doped germanium).

1.6.2.2 Capacitive Transduction: A capacitor consists basically of two metal electrodes separated from each other by an insulating material. The two electrodes are often referred to as the *plates* of the capacitor (condenser), the material which insulates the plates from each other is called the *dielectric* (see Figure 1-10b). There are two basic ways in which capacitive elements are used for transduction.

Fixed Dielectric Method–The two electrodes move toward or away from each other but the dielectric between the plates does not change (air is a commonly used dielectric). One way of doing this is to keep one plate fixed (*stator* plate) and permit the other plate (*rotor* plate) to move toward and away from the stator plate to effect the capacitance changes. This is the principle of the condenser microphone; the movable plate is a thin metallic diaphragm that deflects with incident sound waves; the capacitance changes then represent the incident sound waves.

Similarly, this scheme is used in pressure transducers where the pressure to be measured deflects the diaphragm more or less toward the stator plate; the capacitance change is then proportional to pressure. The rotor plate can simply be attached to a moving object; the capacitance changes are then representative of the motion (displacement). If a seismic mass is attached to the moving plate, the capacitance changes can be made representative of acceleration. If the moving plate is designed to accept, and deflect with, mechanical loading, the capacitance changes are representative of force. Capacitance increases when the plates are brought closer together, and decreases when they move apart from each other. A single set of electrodes is commonly employed; however, the same principle is served when two or more sets of electrodes (with all stators electrically paralleled and all rotors electrically paralleled) are used.

Variable Dielectric Method–Both electrodes are fixed in position; the capacitance changes are effected as the material between the electrodes changes in its *dielectric constant.* This scheme is used primarily in liquid-level sensors. The two electrodes protrude into a vessel so that they get immersed increasingly as the liquid rises in it. The liquid must be essentially nonconductive (i.e., a fairly good insulator); it must also have a dielectric constant different from that of air. As air between the plates is replaced by liquid, the dielectric and, hence, the capacitance changes.

1.6.2.3 Inductive Transduction: An inductor (inductance coil) consists of a winding of insulated wire, wound (when used for transduction) around a ferric core (see Figure 1-10c). The inductor can be so constructed that the coil is wound around an insulating hollow cylinder (coil form) in which the core is free to slide in and out. The core can then be attached to a mechanical element which displaces with a given measurand (e.g., a seismic mass in a spring-mass system for acceleration, a deflecting pressure-sensitive diaphragm, or an object whose motion is to be detected). The self-inductance of the coil will then change as a function of the measurand.

Alternatively, the core can be fixed within the coil and the self-inductance of the coil can be changed by externally-induced changes to the flux field of the coil, e.g., by varying the proximity to the coil of a piece of metal in which eddy currents are generated by the flux, or by varying the proximity to the coil of a piece of ferromagnetic material.

1.6.2.4 Ionizing Transduction: This transduction principle involves changes in ionization current, such as that flowing through a gas between two electrodes across which a potential is applied (see Figure 1-10d). This principle is commonly used in nuclear radiation detectors; the specially selected gas and the electrodes are sealed in an envelope provided with a thin metal "window" through which particles and rays of interest enter. The gas will be ionized proportionally to the intensity and type of radiation, and the ionization current will change accordingly. This principle is employed in ionization counters, Geiger counters, and gas-flow counters, which are all classified as radiation detectors.

In a different usage, a source of electrons is built into the same envelope, e.g., a heated filament of tungsten wire. The ionization current will then vary with the density and type of gas admitted into the envelope through an opening (as indicated by the dotted lines in the envelope, in Figure 1-10d). If this device is then connected, with appropriate plumbing, to a vacuum system, the device will serve as a vacuum gage. A phenomenon similar to ionization also occurs in certain semiconductors that are used as radiation detectors.

1.6.2.5 Derived Non-Self-Generating Transduction Elements: Several widely-used types of transduction elements are derived from the basic resistive and inductive elements described above.

Potentiometric transduction is derived from the resistive element. It employs a resistor of the wire-wound, conductive-film, or conductive plastic type, with at least a portion of the element left uninsulated so that a movable contact can slide along the element (see Figure 1-11a). The amount of resistance between the contact *(wiper arm)* and one end of the resistor changes with motion of the wiper arm. The wiper arm is attached or mechanically linked to a deflecting sensing element or to an object whose motion (displacement) is to be measured.

A source of excitation voltage is connected across the whole resistor. The voltage developed between wiper arm and one end of the resistor is then a function of the ratio of the partial resistance (ΔR) to the resistance of the entire resistor *(resistance element)* (R); hence, the output is a voltage ratio. Either ac or dc can be used for excitation. When the wiper is at one end of the element the output voltage will be zero percent of the excitation voltage; when it is at the other end, the output will be 100%, and when the wiper is in the center of the element the output voltage will be 50% of the excitation voltage.

Reluctive transduction is derived from the inductive element. However, two or more coils (or separated portions of one coil) are employed and what changes is not the self-inductance of a coil but the ac output voltage of the coil system, provided that ac excitation is applied to this system. The output voltage changes are due to changes in the reluctance path between the coils. Such changes influence the amount of coupling between the primary and secondary coils in the example shown in Figure 1-11b, the *differential transformer.* As the core is moved from its center position *(null position)*, the coupling from the primary increases to one of the secondary coils while it decreases to the other secondary coil. As a result; the output voltage changes in amplitude as well as in phase. The amplitude indicates by how much the core was moved. The phase indicates in which direction (from the null position) the core was moved. The core can be attached or mechanically linked to a deflecting mechanical sensing element (diaphragm, seismic mass in a spring-mass system, load-cell column, etc.) or to an object whose displacement or motion is to be measured.

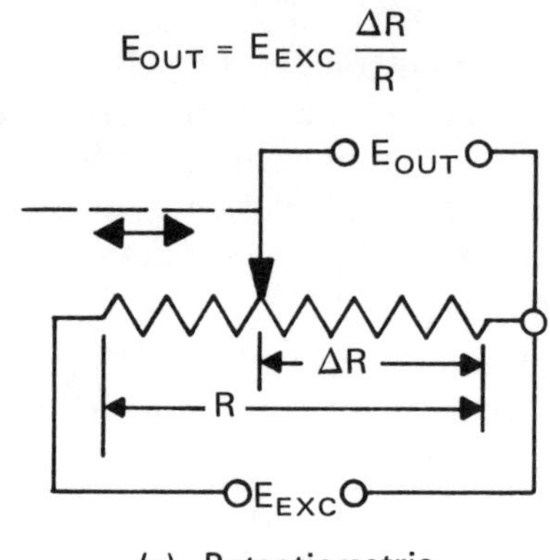

(a) Potentiometric

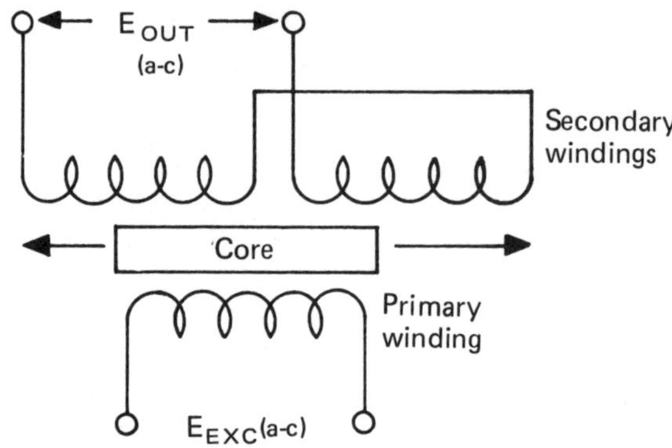

(b) Reluctive (differential transformer)

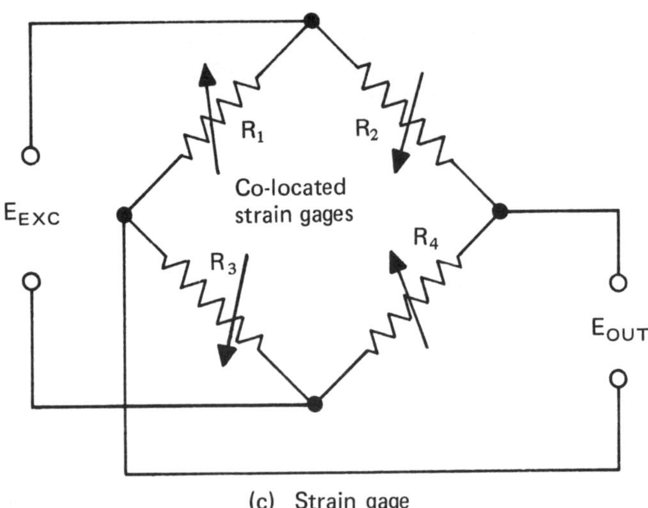

(c) Strain gage

Figure 1-11. Derived passive transduction elements

Another type of reluctive element is the *inductance bridge.* It consists of two inductors connected into a bridge circuit (see Figure 1-12b). A movable core is so arranged that, as it moves, it will increase the inductance in one coil while it decreases the inductance of the other coil. Again, the coil system is excited with ac and the output is a varying ac output voltage. A third type has been used for force transduction. The core is fixed but is made of material which changes its reluctance (and, hence, the coupling between primary and secondary coils) when the core undergoes mechanical strain.

Strain-gage transduction is derived from the resistive element. This type of transduction employs four strain-sensitive resistors (strain gages) which are connected as a Wheatstone-bridge circuit and are so mounted to a mechanical deflecting sensing element that two gages, in opposite bridge arms, will increase in resistance while the other two will decrease in resistance when deflection occurs (see Figure 1-11c). Excitation (E_{EXC}) may be either ac or dc. The output (E_{OUT}) of such a transduction element is four times as large as would be obtained by connecting just a single strain gage in the bridge circuit, shown in Figure 1-12b, as the variable element.

The four gages are matched to each other in resistance, i.e., they have the same nominal resistance. The gages can be of the metal-wire, metal-foil or semiconductor type. Some semiconductor-strain-gage transduction elements have been made out of a single slice of silicon with the strain-sensitive pattern obtained by controlled doping of the silicon; they are known as *integrally-diffused* elements. In some pressure transducer designs such an integrally-diffused element, disc-shaped, is also used as the pressure sensing element; the pressure is applied directly to the other (nondoped) surface of the silicon wafer, which then acts as diaphragm.

Other transduction elements include: the *Hall-effect* element, in which a semiconductor develops a voltage when a transverse magnetic field is applied; the vibrating element, a mechanical element that is set into oscillation at its resonant frequency which will vary as the element is subjected to mechanical stress, so that the element can produce an ac output varying in frequency; the *servo* element in which a mechanical element with a (capacitive, inductive, etc.) transduction element is kept balanced by a torquer or other restoring forcer in a closed-loop arrangement, so that the current required by the torquer or forcer is proportional to the measurand; and the *pyroelectric* element, a self-generating element made of a material that produces a voltage change when it undergoes a temperature change.

1.6.3 Measuring Circuits

The basic passive transduction elements produce outputs in the form of resistance, capacitance or inductance changes (the ionizing element, which provides current changes, is an exception). These outputs must first be converted into a current or (more commonly for biomedical sensors) a voltage that can then be accepted by the signal transmission, telemetry, signal conversion, or display equipment. For this purpose they get connected into a *measuring circuit,* two examples of which are shown in Figure 1-12.

When the element is resistive it can be connected into a *voltage-divider* circuit (Figure 1-12a); a constant-current power supply is often used for such circuits. As the resistance of the variable element changes, the voltage that appears across the fixed resistor will change accordingly, since the variable element changes its portion of the total series resistance constituted by it plus the fixed resistor. Either ac or dc power may be used, depending on the desired form of the output voltage.

The *bridge circuit* is a very popular measuring circuit (Figure 1-12b). A typical bridge circuit contains two fixed resistors whose resistance is identical. When the resistance or impedance of the reference element is identical to the resistance or impedance, respectively, of the variable element, the bridge is balanced, i.e., the output voltage is zero. When both of these elements are resistive the bridge may be excited by either ac or dc. However, when these elements are capacitive or inductive, the power supply must provide an ac voltage having a frequency suitable to make the impedances of these elements large enough to measure easily.

In a laboratory version of such a bridge *(resistance bridge, impedance bridge)* the reference element is variable and provided with a knob and a calibrated dial. The reference is adjusted until its resistance (or impedance) matches that of the "unknown" (variable) element, as indicated by zero bridge output voltage, and the value of the unknown is read off the calibrated dial.

A different method of using a bridge circuit is employed as a measuring circuit for a transduction element. The reference element is selected so that its value (of resistance or impedance) corresponds to that value of the variable element that is indicative of zero measurand (or another value of the measurand that is defined as the bottom of the measuring range); under these conditions the bridge output voltage will be zero. As the value of the variable element changes, due to measurand changes, the bridge output will increase proportionally.

Among other measuring circuits are oscillator circuits, in which the transduction element is inserted as the frequency-controlling element,

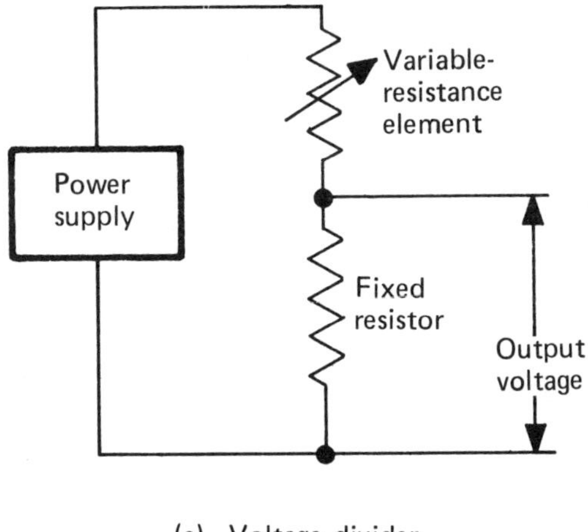

(a) Voltage divider

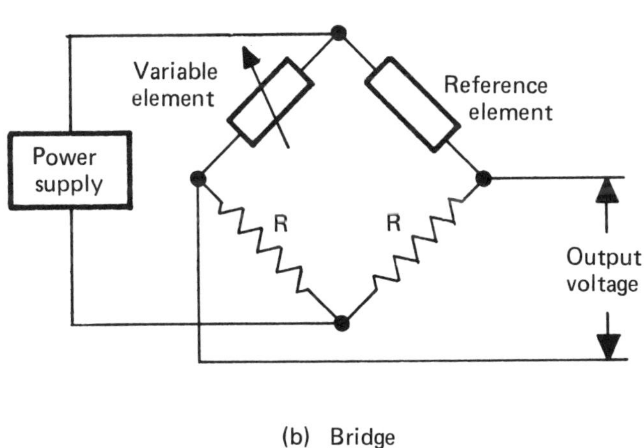

(b) Bridge

Figure 1-12. Typical measuring circuits

and time-delay circuits, in which the transduction element controls the time constant.

1.7 OTHER SIGNIFICANT SENSOR DESIGN FEATURES

Besides measurand, transduction principle, and sensing element there are certain other features of sensor's design that enter into the description or categorization of a sensor. Sensors may be designed for underwater use, or for use with specific, including highly corrosive, fluids. They may be of the type that gets immersed in a liquid or mounted to the outside surface of a pipe or tank. Biomedical sensors may be implantable, intrusive (to varying degrees) or nonintrusive. Some sensors have a measuring circuit and possibly also some signal-conditioning and excitation-conditioning circuitry integrally packaged within the housing of the sensor. Sensors can also be designed to be taped-on, cemented-on, strapped-on, or hand-held. Such features are often important criteria in sensor selection.

1.8 SENSOR PERFORMANCE CHARACTERISTICS

There are three major categories of sensor performance characteristics: static characteristics, dynamic characteristics and environmental characteristics. These are explained below. In addition, reliability, safety and, for applications requiring this, biocompatibility must be considered.

1.8.1 Static Characteristics

The relationship of the output of a sensor to the measurand *(transfer function)* can be described by a *theoretical curve.* The shape of this curve depends on sensor type, design, and materials, but it can be predicted very accurately. A theoretical curve can be linear (straight line) or nonlinear (see Figure 1-13). Examples of nonlinear curves are the bow curve (which could also have a slope opposite to that illustrated) and the S-curve shown in this figure. A convenient way of showing such a curve is by a plot of *output,* in percent of full-scale output (% FSO), vs measurand, in percent of *range,* since this provides a general representation. The output may be a voltage (e.g., 0-5 V full scale), a current, a resistance, capacitance, or a frequency. The range is that range of measurand values over which the full-scale output is provided by the sensor. The range can start at zero measurand (e.g., a flow of 10 liters/minute) or at some other value (e.g., 30°C).

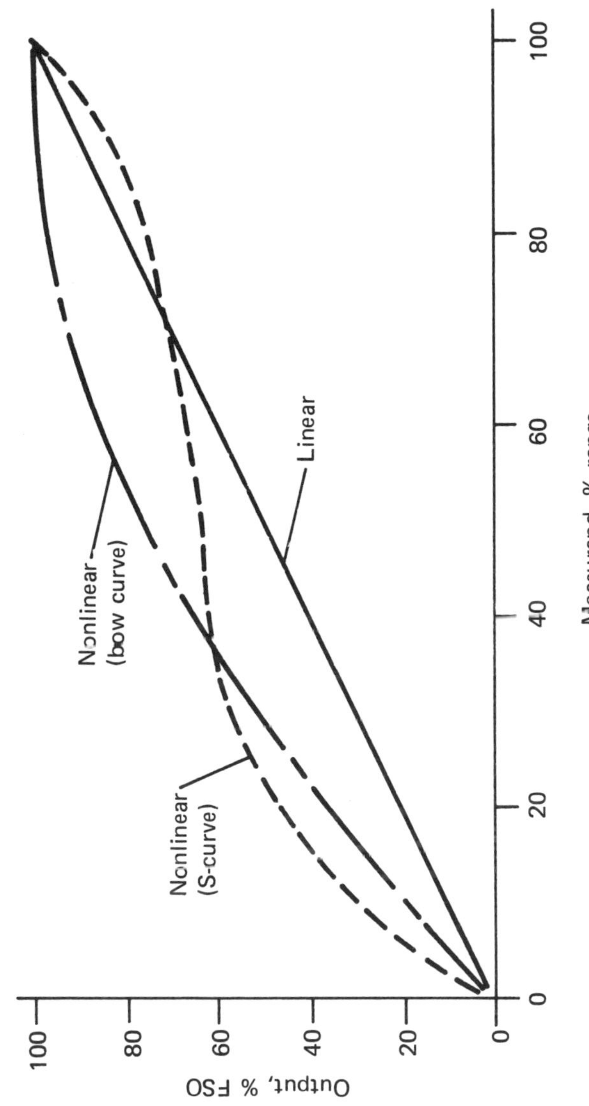

Figure 1-13. Typical sensor output-vs-measurand relationships

If the sensor were an ideal device, the output for any given value of measurand within the range would always be the same output value, regardless of how many times the measurement is made, whether the measurand was increasing or decreasing, or under what conditions the measurement was made.

The output readings of a real transducer, however, will deviate from the theoretical curve due to various factors such as manufacturing variations, behavior of materials, aging, and design tolerances. The output readings will then indicate a value of measurand that differs from the true value. The difference between indicated and true value is the *error* of the sensor. These errors can be determined by a (static) *calibration* of the sensor, during which known values of measurand (known by virtue of being traceable to a government-approved standard) are applied to the sensor and output readings are recorded.

It may then be observed that output readings for the same values of measurand will be slightly higher when the value is approached with decreasing measurand than with increasing measurand; this is due to *hysteresis* which is often noticed particularly when the sensor employs a mechanical (deflecting) sensing element. When a measurand value is approached repeatedly, with the measurand always varying in the same direction (e.g., always with increasing measurand) the differences in output readings indicate the *repeatability* of the sensor.

When the theoretical curve of the sensor is linear and a plot of output readings indicate deviations from a (specified type of) straight line, these deviations indicate the *linearity* of the transducer. Alternatively, the deviations of output readings obtained repeatedly and with the measurand varied increasingly as well as decreasingly, can be plotted as an envelope, showing deviations from the theoretical curve. This envelope is the *error band* of the sensor. Errors as well as error bands can be conveniently expressed in terms of % FSO, e.g., hysteresis–0.8% FSO, repeatability–within 0.5% FSO, linearity–within ±1.2% FSO or error band–+1.5% and −0.5% FSO.

There are other forms of expressing a transfer function and errors. *Sensitivity* is the slope of a linear theoretical curve and is expressed in units of output per unit of measurand, e.g., 5 ohms/°C. Errors can then be expressed in terms of *sensitivity shift* (changes in the slope) and *zero shift* (changes in the output obtained at the bottom of the measurand range). Repeatability, in terms of deviations from a calibration curve furnished with the sensor, is the most significant of these error (or "accuracy") characteristics.

1.8.2 Dynamic Characteristics

Dynamic characteristics describe the behavior of a sensor in response to a step change in measurand or in response to rapid fluctuations in the measurand. A microphone, for example, has to respond to sound waves varying in frequency from a few hertz to hundreds, sometimes thousands of hertz. The *frequency response* (see Figure 1-14) must then be established. The *frequency range* of the sensor is the frequency response within a specified tolerance band from the 100% output/measurement ratio level.

When a sensor experiences a step change in measurand (e.g., when a thermistor thermometer is inserted orally or rectally), the sensor will respond more or less gradually to this step change (see Figure 1-15). The output will rise (or fall) from its initial value to its final value. If the time at which the step change occurs is defined as "zero" (e.g., zero seconds, as in the illustration), the time at which the output reaches 63% of its final value is the *time constant.* The time at which the output reaches a specified percentage (e.g., 95% as illustrated) is the *response time* of the sensor, always accompanied by statement of the percent output change, e.g., 95% response time. The time elapsed between a low and a high percent output change, e.g., 5-95%, or (as illustrated) 10-90%, is the *rise time,* always accompanied by a statement of both percentages, e.g., 10-90% rise time.

1.8.3 Environmental Characteristics

The thermal, vibration, shock, etc., environments in which the sensor must operate within specified tolerances may differ from the *room conditions* under which the sensor was calibrated. Such *operating environmental conditions* must be known and their effects on the behavior of the sensor must be established. The effects can often be additional errors (deviations from the theoretical curve or the calibration curve).

The most significant of these, for biomedical sensors, is *temperature error,* the differences in output readings at a high or low temperature from the same readings taken at room temperature. When sensitivity was specified, such errors are often expressed as *thermal sensitivity shift* and *thermal zero shift.* Such errors are usually shown in % FSO but can sometimes be seen expressed as sensitivity shift per degree of temperature.

Vibration errors, acceleration errors, and ambient-pressure errors (errors due to changes in the ambient barometric pressure) must also be considered in certain sensor applications. Errors or erratic behavior can also be induced by nearby strong magnetic fields or by nuclear radiation.

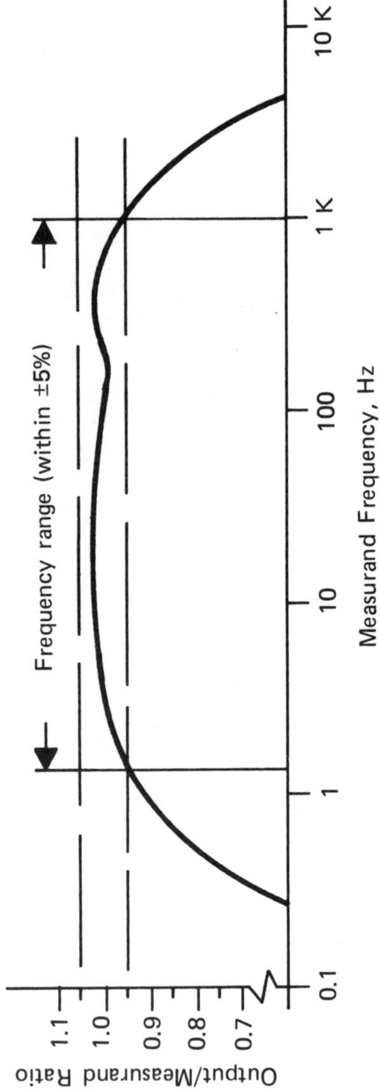

Figure 1-14. Frequency response

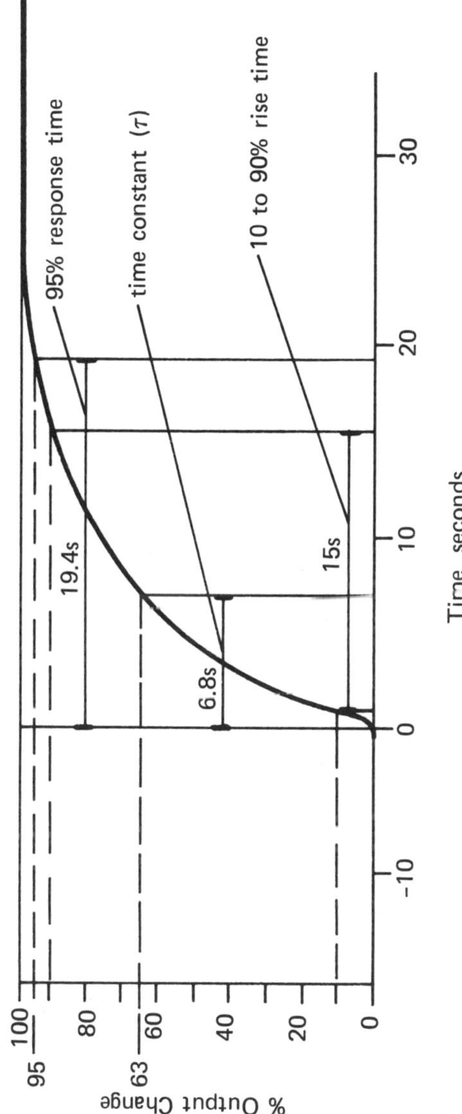

Figure 1-15. Example of time response to step change in measurand

In many cases the *nonoperating* environmental conditions must also be known, together with their effects on a sensor. The sensor must be able to operate within specifications after, but not during, exposure to such conditions. An example would be a high sterilization temperature.

2. Physiological Fundamentals

2.1 STRUCTURAL ORGANIZATION OF THE BODY

All living matter is composed of *cells* and cell products. The cell consists of protoplasm surrounded by a membrane. Cells and intercellular substances combine to form *tissue* of which there are four types: epithelial (covering the internal and external body surfaces), connective, muscle and nervous. A group of tissues serving a common function combine to form an *organ,* a single structure such as the heart, lungs, etc. Cells, tissues and organs combine to form a *system.* The major systems of the body are the skeletal, muscular, nervous, cardiovascular (circulatory), respiratory, endocrine, digestive, urinary and reproductive system. The skin is also considered a system (integumentary system). The special senses (hearing, vision, etc.) are considered part of the nervous system and the lymphatic system a part of the circulatory system.

The status and operation of physiological systems can be inferred from signals provided by them *(bioelectric potentials)* and from physical-quantity measurements made in or on the systems. For the purpose of this book it is useful to consider the engineering analogs of these systems, even at the risk of some oversimplification.

The *skeletal system* is a structural system which lends itself to determinations of such physical parameters as strength, elasticity, displacements and force. The *muscular system* is an electromechanical system. Signals within this system can be sensed, amplified and displayed *(electromyogram).* Sometimes these signals are obtained in response to an externally-applied electric stimulus. The *nervous system* is an electrical sensing and control system. Signals *(action potentials)* at many points in this system can be sensed, amplified and displayed. The electrical activity of the brain can also be measured and recorded *(electroencephalogram)*.

The *respiratory system* is a pneumatic system. Most of its essential physical parameters can be measured by means of sensing devices. The *cardiovascular system* is an electrically-controlled hydraulic system which includes the pumping functions of the heart and the flow of blood through the circulatory system. It lends itself to monitoring of such physical quantities as (blood) pressure and flow rate, as well as of the electrical potentials associated with the operation of the heart *(electrocardiogram)*.

The *integumentary system* (the skin) can provide information about some physiological functions (including psychologically-induced phenomena) through variations in point-to-point electrical resistance measurements. Temperatures at specific points on the skin or of skin surface areas can be determined by direct or noncontacting measurements.

Physical displacements of organs or other portions of the body and their derivatives (velocity, acceleration) can be sensed by various techniques. Dimensional information can be obtained by imaging techniques, using various types of penetrating illuminations such as X-rays, gamma-rays and ultrasound, and by appropriately located sensors responding to each type of illumination.

Since most biomedical measurements involve an interaction between electronic equipment and the human body, it is of extreme importance that no potential shock hazards exist in the use of such equipment. Electrical safety–proper isolation of elements and proper grounding–has come under close scrutiny and has become the subject of extensive and thorough standardization efforts.

The systems of the body in which most of the existing biophysical sensing and analyzing devices are used are the cardiovascular, nervous and muscular, and respiratory system. The basic concepts of these systems, primarily those related to biomedical measurements, are briefly defined below.

2.2 CARDIOVASCULAR SYSTEM

The *cardiovascular system* consists of the heart, the blood, and the blood vessels through which the blood flows in its circulation through the body.

The *heart* is an organ whose function it is to pump blood through the body. The major pumping action is performed by the contraction, followed by the relaxation, of two cavities *(ventricles)*. Each ventricle obtains the blood to be pumped from a fore-chamber *(atrium)* which acts as a reservoir and also performs the minor pumping action of forcing its contents into the ventricle. Specially-constructed valves at the inlet and outlet of each ventricle act as check-valves to assure one-way flow of blood. The right ventricle pumps deoxygenated blood to the lungs. The left ventricle pumps oxygenated blood into all other parts of the body. The oxygenation occurs in the lungs.

The *blood vessels* are the passive portions of the cardiovascular system (Latin: *cardium* = heart, *vas* = vessel, *vasculum* = small vessel). The *arterial* blood vessels carry blood *from* the heart, whereas the *venous* blood vessels carry blood *to* the heart. The blood vessels in both systems are largest in diameter nearest the heart, then branch out with decreasing diameters throughout the body in a manner comparable to the root structure of a tree. The larger-diameter vessels *(arteries, veins)* eventually branch out into vessels of much smaller diameter *(arterioles, venules)*, which, in turn, branch out into a multitude of (arterial and venous) *capillaries* whose ends are in sufficiently close proximity that circulation *(systemic circulation)* takes place from the arterial system through the capillaries to the venous system.

Blood is the fluid pumped by the heart through the blood vessels. It carries nutrients and fuel to all parts of the body. It also picks up waste material and carries it to specific parts of the body. It consists of red blood cells (*erythrocytes,* about 5 x $10^6/mm^3$ of blood), white blood cells (*leucocytes,* about 5 x $10^3/mm^3$ of blood), cell fragments involved in clotting (*thrombocytes,* or *platelets,* 5 x $10^5/mm^3$ of blood) and plasma (which is about 91% water, the rest being protein, salts, nutrients, and clotting particles).

Blood pressure is the (gage) pressure of the blood in the heart and in the circulatory system. Typical ranges of pressure are shown in the hydraulic-equivalent simplified diagram of Figure 2-1. Rhythmic pressure fluctuations in the system are due to the pumping action of the heart. At any given point in the system, the *systolic pressure* is the peak pressure resulting from ventricle contraction, whereas the *diastolic pressure* is the static pressure prevailing during ventricle relaxation. The *pulse pressure* (not shown in the diagram) is the difference between the systolic and diastolic pressure at a given point.

Blood flow is the volumetric flow rate of blood, at any point in the circulatory system. It is maximum (about 3.5-5 liters per second) in the pulmonary artery and in the aorta (the largest blood vessel of the arte-

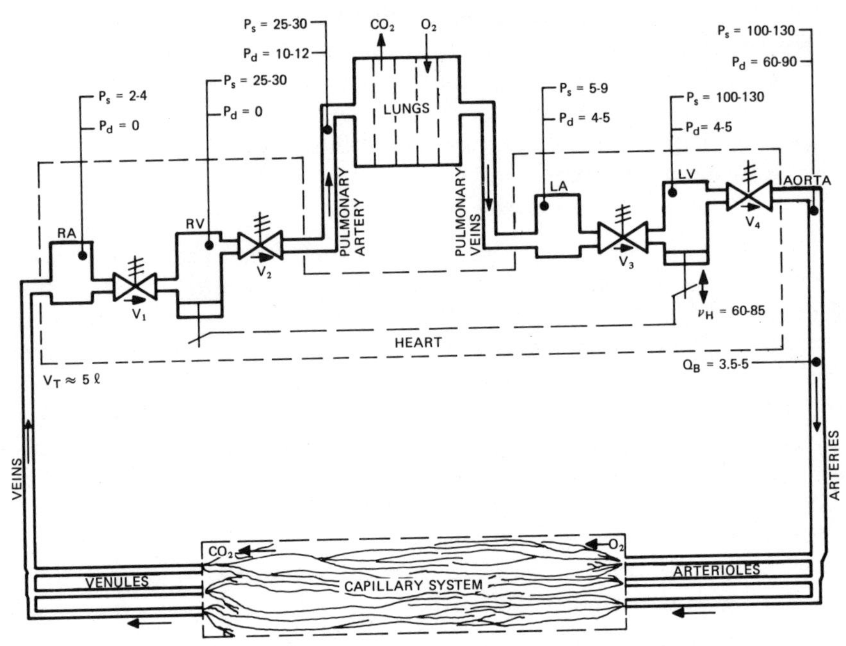

LEGEND:

P_s - Systolic pressure (mm Hg)
P_d - Diastolic pressure (mm Hg)
ν_H - Heartbeats (per min)
Q_B - Blood flow rate (ℓ/min)
V_T - Total volume of blood in body
RA - Right atrium
RV - Right ventricle
LA - Left atrium
LV - Left ventricle
V_1 - Tricuspid valve
V_2 - Pulmonary valve
V_3 - Mitral valve
V_4 - Aortic valve

Figure 2-1. Simplified hydraulic diagram of cardiovascular system

rial system), as shown in Figure 2-1. The flow rate at these points is also referred to as *cardiac output.* The blood flow rate then diminishes as the blood flows through the vessels, increasing in number and decreasing in their diameter, and is barely perceptible in the capillaries, over a million in number and approximately 8 μm in diameter. The total volume of blood in the body (about 5 liters) thus circulates through the system about once every minute.

The *heart-beat* is one full contraction-relaxation cycle of the heart. The typical heart-beat rate for an adult human, at rest, is 60-85 per minute (see Figure 2-1). *Stroke volume* is the cardiac output divided by the number of heart-beats per minute. The *pulse* is the sequential expansion and contraction of the walls of the arteries, usually expressed as the number of these dimensional variations per minute.

Neuromuscular cardiovascular control is the control of the actions of the heart and of the blood vessels, actuated by muscles and initiated by stimuli transmitted by nerves (Figure 2-2).

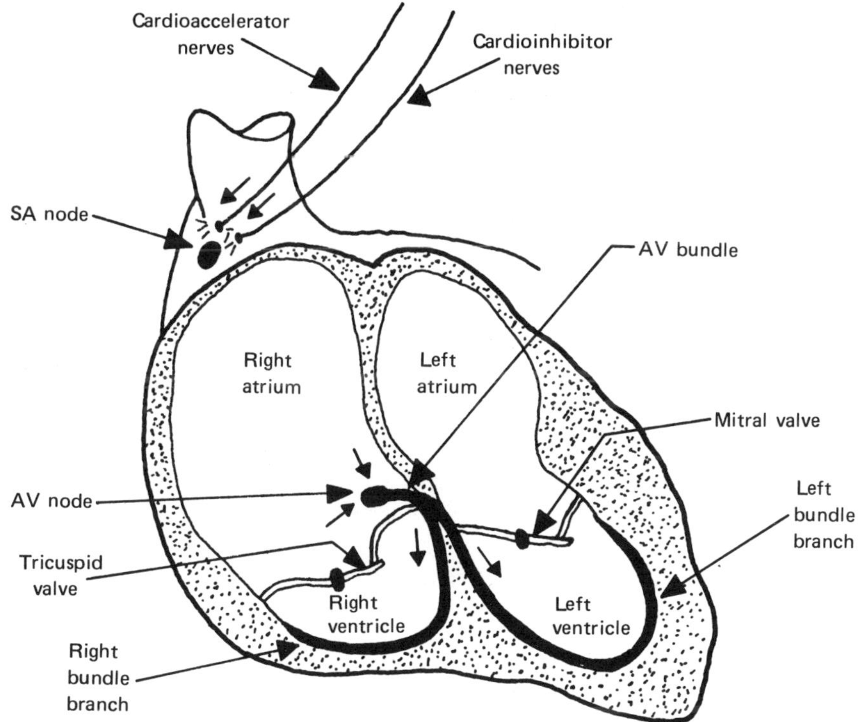

Figure 2-2. Heart stimulus conduction schematic diagram

The conduction system of the heart starts at the *sinoatrial* (SA) *node* which is located in the wall of the right atrium. This node is also the connecting area for two nerves originating in the brain: the *cardioaccelerator nerve* (to make the heart beat faster) and the *vagus nerve* (to make the heart beat slower). The impulse at the sinoatrial node first excites the two atria and causes them to contract. This excitation of the atria causes a stimulus at the *atrioventricular* (AV) *node* located in the wall between the two atria (the interatrial septum). This stimulus generates an impulse through bundles of modified muscle fibers to cause contraction of the ventricles.

The impulse travels first through the *atrioventricular bundle (bundle of His)* to delay contraction of the ventricles until they have had time to fill, then through the *right bundle branch* and the *left bundle branch* (to activate the right and left ventricle, respectively) which then connect through special fibers *(Purkinje fibers)* to the cardiac muscle structure (the *myocardium)*, causing the ventricles to contract and pump blood.

The brain also originates control of blood vessel diameters, an increase in diameter *(vasodilation)* or a decrease in diameter *(vasoconstriction)*. This control of heart and blood vessel actions is influenced by signals to the brain from sensory receptors, including the *baroreceptors* (which respond to pressure variations) and the *chemoreceptors* (which respond to variations in chemical composition) located in the arteries.

2.3 NERVOUS AND MUSCULAR SYSTEMS

The *nervous system* is subdivided into three parts: the *central nervous system* (brain and spinal cord), the *peripheral nervous system* (sensory input nerves which carry information to the brain, cranial motor nerves, originating at the brain, and spinal nerves, originating in the spinal cord), and the *autonomic nervous system* (which transmits impulses to internal organs, glands and blood vessels).

Nerves carrying information to the brain are called *afferent* nerves, whereas nerves carrying information from the brain are called *efferent* nerves. The autonomic nervous system is entirely an efferent system. The central and peripheral nervous systems contain efferent as well as afferent nerve fibers.

The *brain* (Figure 2-3) comprises the forebrain *(cerebrum, thalamus* and *hypothalamus)*, the brain stem (*midbrain*, which connects the cerebrum with the pons and cerebellum, the *pons* and *medulla*, often referred to as medulla oblongata) and the hindbrain *(cerebellum)*. The cerebrum is longitudinally divided into two *hemispheres* and each of these is divided into four *lobes* (frontal, parietal, temporal, and occipital). The two hemispheres are connected by a bridgelike structure *(corpus callosum)*. The outer surface of the cerebrum (within the skull) is the *cerebral cortex*, gray matter arranged in convolutions (folds).

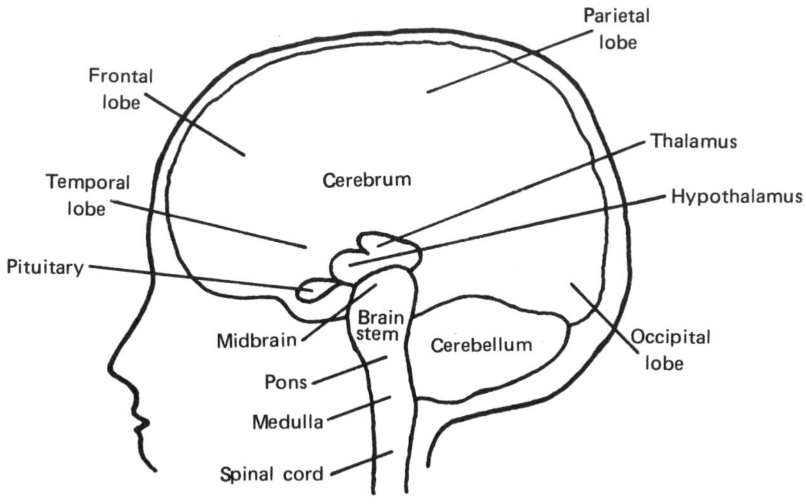

Figure 2-3. Major divisions of the brain

A *nerve* is a bundle of nerve fibers which can be sensory (afferent) or motor (efferent) fibers. The human body contains 12 pairs of cranial nerves and 31 pairs of spinal nerves.

The *neuron* (see Figure 2-4) is the basic unit of the nervous system. It is a single cell, consisting of a *cell body, dendrites* which increase the receptive surface of the cell, and an *axon*, a long cell extension which transmits impulses.

Sensory (afferent) neurons conduct impulses from a *receptor* (an end organ which responds to stimuli such as touch, temperature, pain, etc.) to the central nervous system. *Motor* (efferent) neurons conduct impulses from the central nervous system to *effector* organs such as muscles or glands. A third category are the *internuncial* (association) neurons within the central nervous system. They transmit impulses between neurons. The transmission path is from the axon of one neuron to the cell body or dendrite of another neuron via a (chemical) interconnection point *(synapse)*.

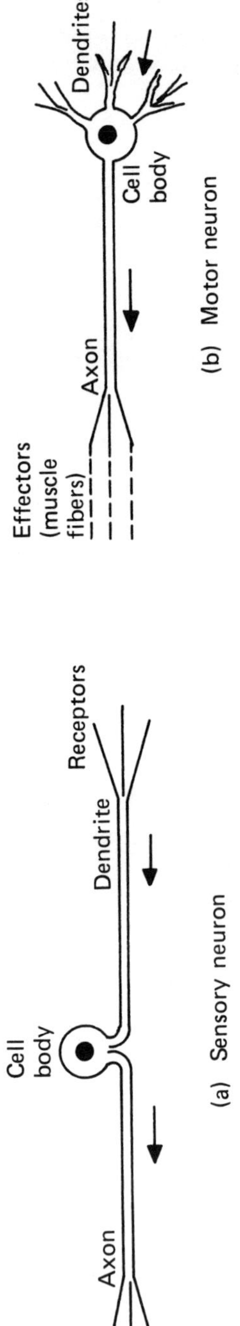

Figure 2-4. Neurons—schematic representation

Many types of neurons are surrounded by a concentric layer of a fatty insulating material *(myelin)* which is typically interrupted to form short segments. These interruptions of the coating of *myelinated* neurons, which serve to increase the transmission velocity along the neurons, are called *nodes of Ranvier*. Many myelinated neurons are further covered by a very thin outer insulating layer *(neurilemma)*. The human nervous system contains about 10×10^9 neurons.

A *nerve impulse* is a (self-propagating) electrochemical potential which travels along the (quasi-cylindrical) neuron-membrane surface (see Figure 2-5). The polarity is taken as that of the inside of the membrane with regard to the outside of the membrane. The potential of the inside, during the resting and recovery modes, is about −70 mV. In response to a stimulus, however, a polarization/depolarization potential occurs which reaches about +30 mV on the inside *(action potential)*. The nerve impulse travels along the length of a nerve fiber at a velocity of between 0.5 and 110 m/s, depending on the type and characteristics of the fiber.

The amplitude and waveshape of the nerve impulse is constant with any variations in the magnitude of a stimulus above a certain threshold. The parameter affected by stimulus increases is, instead, the repetition rate of nerve impulses. Action potentials do not travel across a synapse to another neuron but cause the generation of a new action potential in the postsynaptic neuron.

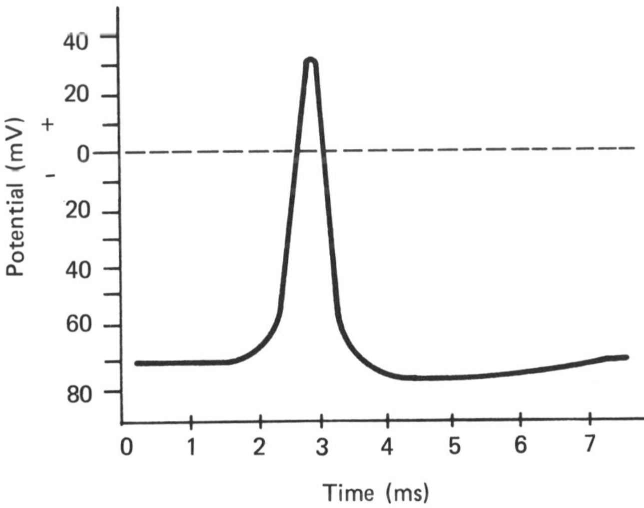

Figure 2-5. Nerve impulse

The *muscular system* provides movement and contraction. It comprises *striated muscles* (voluntary action), *smooth muscles* (involuntary action) and *cardiac muscles* (combination of striated and smooth muscle characteristics, involuntary). The striated appearance is caused by alternating light and dark bands of protein fibrils.

A *muscle* consists of a group of *muscle fibers* which, in turn, are made up of *muscle cells.* The human body contains roughly 250 x 10^6 muscle cells.

A *motor unit,* the entity providing contraction, consists of a group of muscle fibers all of which are innervated by the terminal filaments of a motor axon (see Figure 2-6). The strength of contraction of a motor unit depends on the repetition rate of nerve impulses. The strength of muscular contraction also depends on the number of motor units which are stimulated. The degree of fine control of whole muscles, however, improves with a decreasing number of muscle fibers in a motor unit.

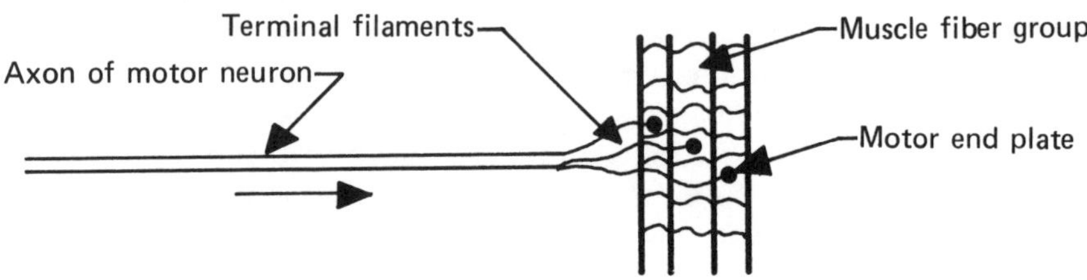

Figure 2-6. Motor unit

A *tetanic* contraction is a prolonged contraction with increase in strength. Muscle *tone (tonus)* is a state of readiness of muscles effected by stimulation of motor units in rotation.

The *refractory period* of nerve or muscle cells is the period following a stimulus application (which results in an action potential) during which the cell will, first, not respond to any further stimuli (absolute portion) and, next, during which increasing stimulation is required to produce an action potential (relative portion).

A *reflex arc* is the basic unit of integrated neuromuscular activity. A typical reflex arc starts at a sensing organ, at which a potential is generated, continues with an afferent neuron, a synapse and an efferent neuron, with action potentials along the afferent neuron creating other action potentials in the postsynaptic efferent neuron, resulting in a motor end-plate potential at a *myoneural* (muscle/nerve) junction at which the appropriate muscular action is effected by muscular action potentials.

2.4 RESPIRATORY SYSTEM

The function of the respiratory system is to supply oxygen to the body and to dispose of carbon dioxide. This gas exchange takes place in the *alveoli,* the air cells of the lungs (human lungs contain about 3×10^8 alveoli). The passage of air into the lungs starts at the nose and/or mouth, then continues through the *pharynx,* the *larynx* (which also contains the vocal cords) and the *trachea* (windpipe) which branches into the two *bronchi,* one leading to the right lung, the other to the left lung. These main bronchi divide into a tree-like structure in each lung. The numerous small-diameter branches of this structure are the *bronchioles* which eventually connect to the alveoli (see Figure 2-7). During *expiration* the air flow uses the same elements in the reverse sequence as that described for *inspiration,* above.

Breathing (external *respiration*) is accomplished by two sets of muscles which sequentially increase and decrease the volume of the *thoracic cavity:* (a) the muscles of the *diaphragm* whose up-and-down motion causes thoracic-cavity volume changes (*abdominal* or *diaphragmatic* breathing) and (b) the muscles of the rib cage whose upward-and-outward, then downward-and-inward motion effects the required volume changes of the thoracic cavity (*costal* breathing).

Respiration is controlled from inspiratory and expiratory centers in the brain's medulla and pons. A number of factors affect respiration through these centers, e.g., respiration is increased by: increased CO_2 partial-pressure (pCO_2) in the arterial blood and decreased pO_2 (up to a point); a low arterial pH; increased body heat; certain reflexes from joints; and sensory stimuli such as pain or sudden cold.

Lung volume can be expressed for various conditions of normal, quiet respiration and for a maximum-effort inspiration and expiration (values, where shown, are typical and approximate):

(a) normal breathing–the *tidal volume* (500 cm^3) is the difference between the *end inspiratory* and *end expiratory levels;* it is the volume of gas inspired or expired in a resting mode.

(b) maximal-effort breathing–the *inspiratory reserve volume* is the additional volume of air, above the end inspiratory level, whereas the *expiratory reserve volume*

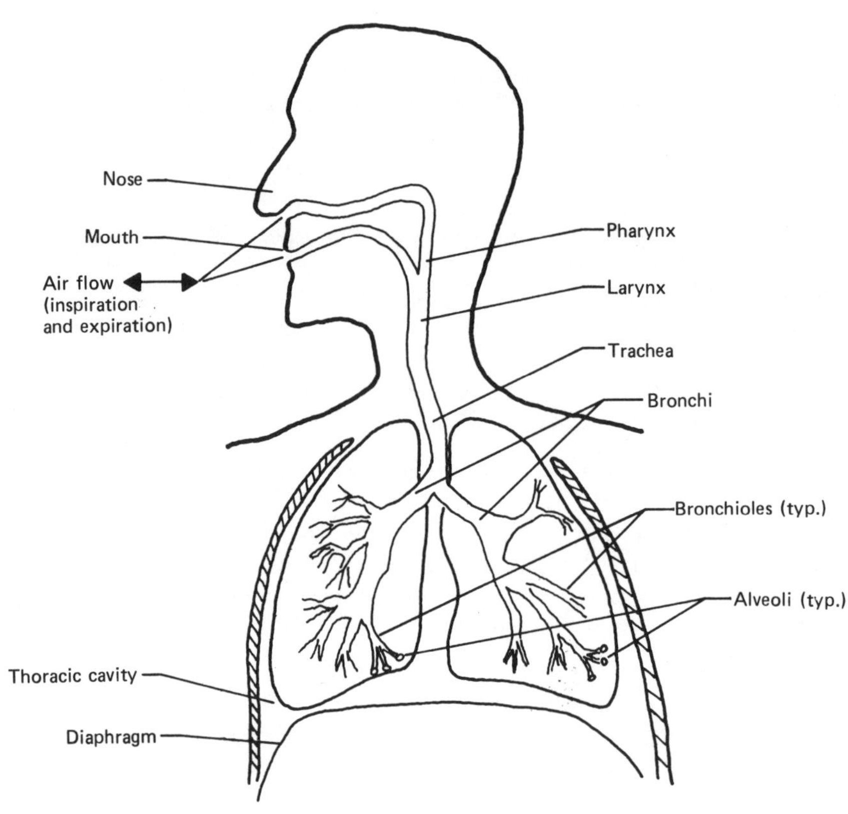

Figure 2-7. Basic elements of the respiratory system

is the volume reduction below the end expiratory level; the volume of air remaining in the lungs after a maximal-effort expiration is the *residual volume* (1100 cm^3).

(c) the *vital capacity* is the difference between the maximal inspiratory level and the residual (volume) level; vital capacity (4500 cm^3) is also the sum of tidal volume (tidal air), inspiratory reserve volume (complemental air) and expiratory reserve volume (supplemental air).

(d) the sum of vital capacity and residual volume is the *total lung capacity* (5600 cm^3); the sum of tidal volume and inspiratory reserve volume is the *inspiratory capacity;* the sum of residual volume and expiratory reserve volume is the *functional residual capacity* (FRC).

(e) the *forced expiratory volume* (FEV) is the maximal-effort expiration over a specified amount of time; *forced vital* capacity is the FEV attained during a minimum amount of time (total expiration as quickly as possible after maximal-effort inspiration); the maximum air that can be inspired and expired over a sustained period (e.g., 20 s) is the *maximal breathing capacity* (or *maximal voluntary ventilation*) 150 ℓ/min).

Lung *compliance* is the change in lung volume per unit change in lung (intrathoracic) pressure, usually taken on the increasing cycle; the relationship is nonlinear.

2.5 INTEGUMENTARY (SKIN) SYSTEM

Basal skin resistance (BSR) is the ohmic baseline resistance between two points on the skin.

Galvanic skin response (galvanic skin resistance, GSR) is the dynamic (decreasing) variation of skin resistance between two points on the skin in response to a stimulus.

Skin potential is a potential difference between two points on the skin (typically forearm and palm abut 50 mV); this potential varies with emotional changes, as does GSR.

3. Cardiovascular System Measurements

3.1 BLOOD PRESSURE

Blood pressure, and its variation with time, is probably the most frequently measured biomedical quantity (besides body temperature). It is measured by direct or indirect methods. *Direct methods* require the puncturing of a major blood vessel so that the blood stream, itself, can be coupled to a pressure measuring system. The blood pressure can be measured either at the point of puncture, or, by means of a specially designed thin tube inserted at the puncture point *(catheter),* at other points in the cardiovascular system including the heart (see Figure 2-1). *Indirect methods* are directed at measuring some outward manifestation of blood pressure which can be correlated to the pressure in the blood stream.

The *sphygmomanometer* is the most widely used device for measuring blood pressure by an indirect method. Its use is based on the *Riva-Rocci-Korotkoff* method. Its basic components are an inflatable cuff, a means for inflating and deflating the cuff, a means for indicating the pressure in the cuff, and a means for listening to (or sensing and displaying) sounds in a blood vessel (see Figure 3-1).

The pneumatic cuff (introduced by Riva-Rocci in 1896) is used to compress an artery to the point where blood flow stops. It is usually placed around the upper arm to determine the pressure in the brachial artery. This point in the cardiovascular system is at a pressure essentially identical to aortic pressure, but much more easily accessible than the aorta.

The sound-detection device (usually a stethoscope) is held against

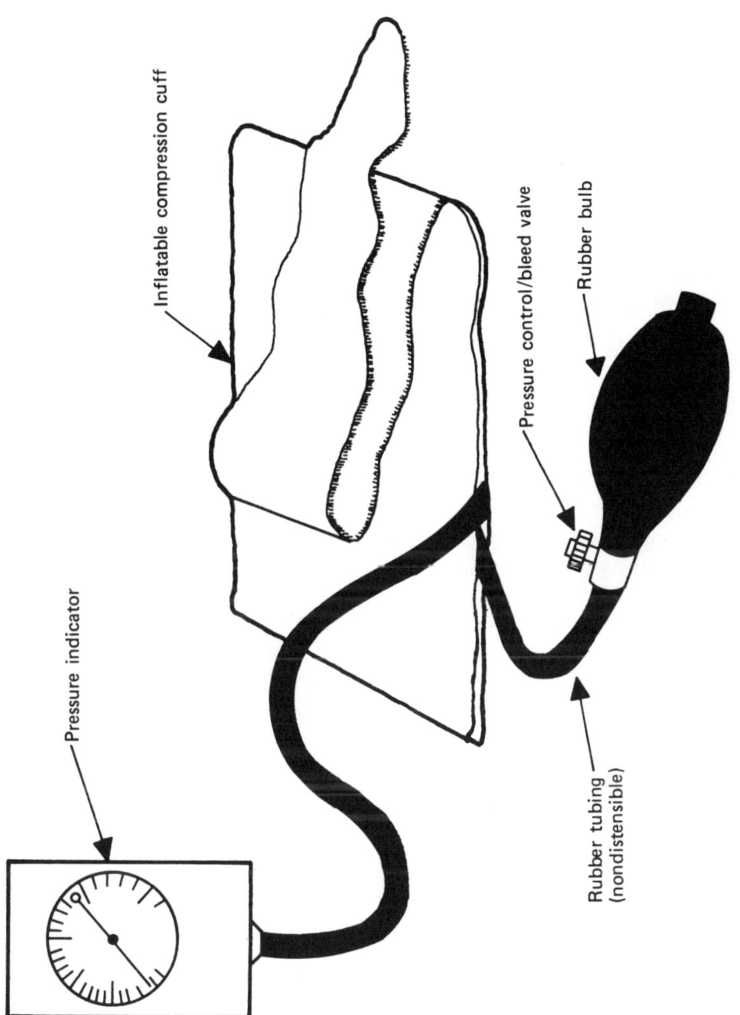

Figure 3-1. Sphygmomanometer

the artery at a point immediately downstream from the point at which the artery is to be compressed by the cuff. The sounds to be detected are the *Korotkoff sounds* which characterize blood flow when cuff pressure is released, on the basis of a method introduced by the Russian scientist Korotkoff in 1905.

A pressure-balancing technique is used to determine blood pressure with a sphygmomanometer. The cuff is inflated until the artery is completely occluded and blood flow stops. At this point the cuff pressure is at least equal to the systolic pressure in the artery. In order to detect the systolic pressure properly, the cuff is inflated slightly above this pressure. By means of the pressure control valve (bleed valve, vent valve) the cuff pressure is then gradually decreased (see Figure 3-2).

The systolic pressure (P_s) is read at the beginning of phase 1 of the Korotkoff sounds, i.e., when the first sharp thud is heard. As the pressure continues to be decreased the sounds change from sharp thuds to blowing or swishing sounds (phase 2), to thuds softer than those of phase 1 (phase 3), finally to a softer blowing sound that disappears (phase 4). At the end of phase 4 (or the beginning of silence, phase 5) the diastolic pressure (P_d) is read.

The cuff pressure is usually indicated by an aneroid manometer (pressure gage), sometimes a mercury manometer. It can also be measured by a transducer whose output is either displayed on an analog or digital display device, or telemetered.

The determination of blood pressure by listening to the Korotkoff sounds (the *auscultatory* technique) requires a sound detection device to be placed as closely over the artery as possible, at a point immediately below the compression cuff (away from the heart). A stethoscope is most commonly used for this purpose and a trained "ear" is required to interpret the sounds properly. In some blood-pressure sensing devices the stethoscope is replaced by a microphone mounted directly in, or to the inside surface of the cuff. The output of the microphone can then be amplified and made audible by headphones, or it can be conditioned so that a light starts flashing when the systole is reached and stops flashing when the diastole is reached, or it can be fed to a visual-display device such as a cathode-ray tube or a strip-chart recorder.

A number of automatic blood-pressure measuring systems of the sphygmomanometer type have been developed. They typically employ a pressure source, a programmable means for inflating and deflating the cuff at selectable inflation and deflation rates, and a cuff microphone. The outputs of the cuff pressure transducer and of the microphone can be recorded simultaneously on a single display device which has been calibrated for proper correlation of pressure with Korotkoff sounds.

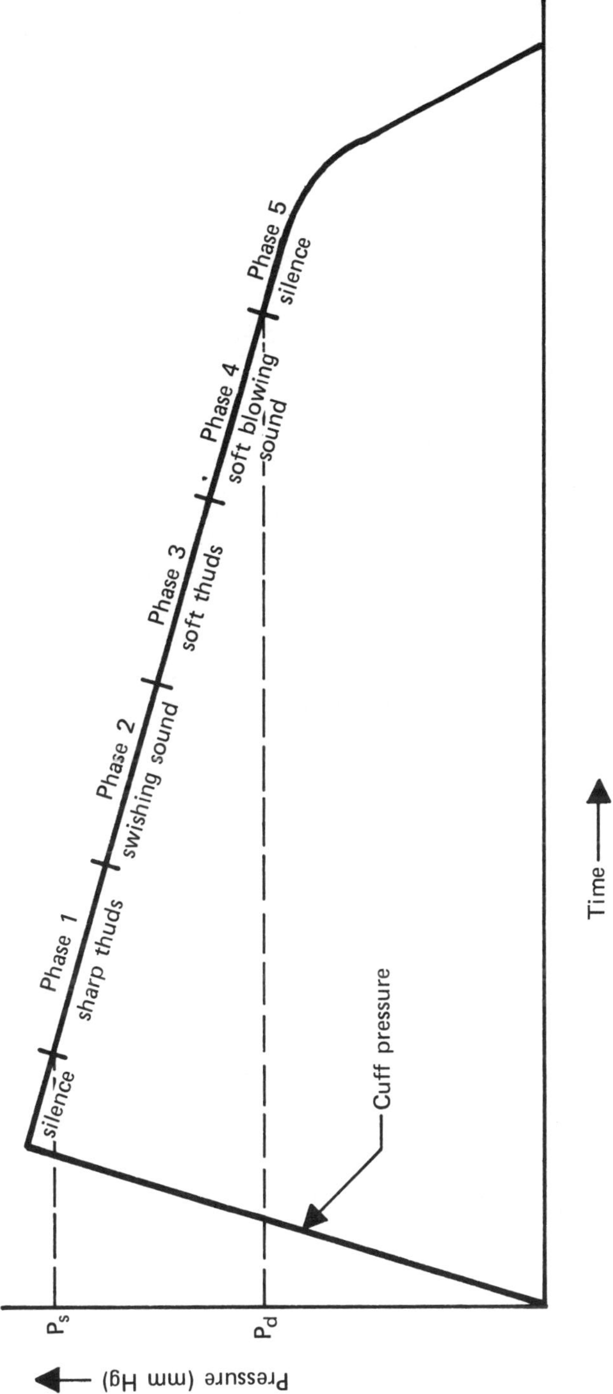

Figure 3-2. Blood pressure determination by Korotkoff sounds

Some other automatic systems have been designed which rely on the pulse, rather than on Korotkoff sounds, for systole and diastole determination. The pulse disappears downstream of the point where the externally-applied pressure balances the systolic pressure. It reappears when the pressure is reduced to the diastolic level. Pulse transducers of various types (e.g., photoelectric, reluctive, strain-gage) have been used.

One such system employs a finger cuff instead of the usual arm cuff. The small pneumatic cuff is placed around a finger and the pulse is detected at the finger tip. Although the pressure levels in the blood vessels of the finger are much lower than in the brachial artery they can be correlated to arterial pressure and reasonably accurate values of systolic and diastolic pressures can be determined if the system is properly designed, applied, and calibrated.

The *ear oximeter,* a photoelectric device for the measurement of percent oxygen saturation of the blood in the ear, has also provided values which can be correlated with arterial pressure. Variations in ear opacity are detected by means of a light beam and a light sensor. One system uses a transparent pressure capsule, pressed against the ear, to perform the function of the pneumatic cuff. Low-frequency light-sensor output signals, which can give erroneous readings when arterial pulse is to be determined, can be reduced by connecting a high-pass filter into the output circuitry.

Direct methods for blood pressure sensing involve surgical procedures. *Percutaneous insertion* is a very minor procedure involving the insertion of a thin tube or hypodermic needle into a blood vessel so that fluid contact can be made with the sensing element of a pressure transducer. The transducer is usually kept external to the body; however, several designs of *catheter-tip transducers* have been manufactured and used successfully. These are ultra-miniature pressure transducers mounted integrally in the tip of a very thin tube *(catheter)* which can be inserted into the blood vessel. Such transducers are also used in *catheterization,* a surgical procudure in which a catheter of suitable length is inserted into a major blood vessel and gently guided to the point at which a pressure measurement is to be taken. This point can be one of the chambers of the heart. Alternatively, the catheter can be a hollow tube connected to a pressure transducer external to the body.

Typical blood pressure transducers, used external to the body, are shown in Figure 3-3 (strain-gage type) and Figure 3-4 (reluctive, LVDT type). Characteristics peculiar to blood pressure transducers are the transparent dome, shown in the illustrations, which permits visual detection of any air bubbles in the blood and the second pressure port

Figure 3-3. Unbonded-strain-gage blood pressure transducer (courtesy of Bell & Howell, CEC Div.)

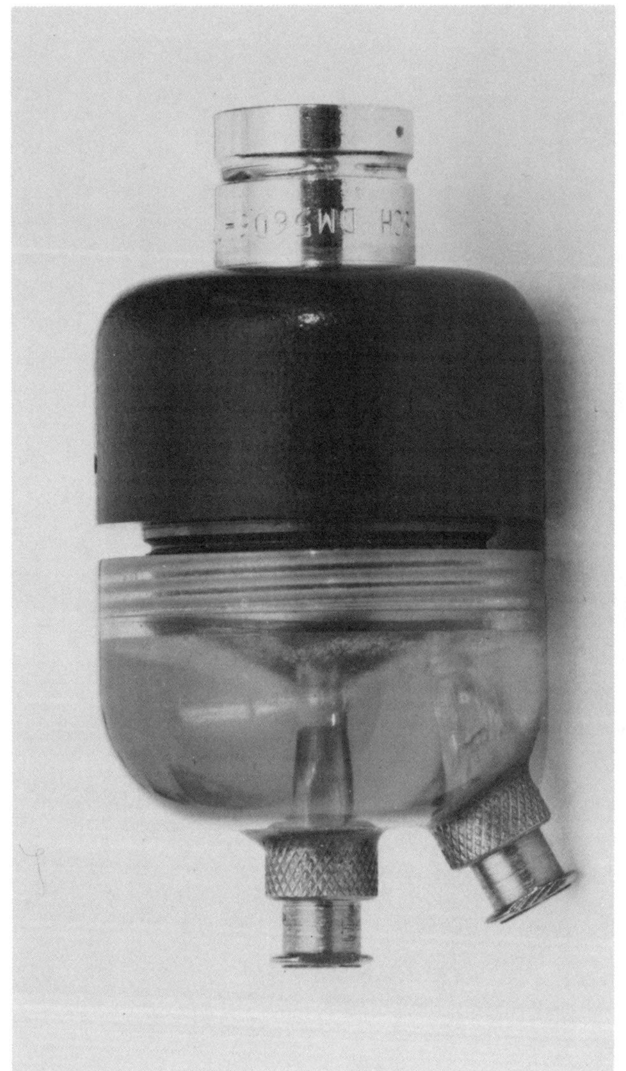

Figure 3-4. Reluctive (LVDT) blood pressure transducer (courtesy of Biotronex Laboratory, Inc.)

which allows flushing of the transducer. Another important characteristic is electrical safety, provided by good (sometimes double or triple) electrical isolation of any current-carrying portions of the transducer from the blood stream (and the patient). An environmental requirement for such transducers is the ability to withstand sterilization (e.g., by autoclave or steam).

Implantable blood pressure transducers have been used in some cases, sometimes with their electrical connecting leads brought out to equipment external to the body, sometimes with provisions for inductive coupling to an external system, and, in a few instances, connected to an implanted miniature telemetry transmitter.

3.2 BLOOD FLOW

The flow of blood through any blood vessel can be measured by invasive methods. The flow through some blood vessels, especially those close to the surface (skin) can also be measured by noninvasive methods. The use of electromagnetic blood flow sensors requires relatively major surgical procedures, whereas the use of thin-film catheter-tip anemometer-type sensors requires relatively minor surgical procedures. Ultrasonic flowmeters have been used for invasive as well as noninvasive methods but lend themselves to noninvasive flow measurements through blood vessels close to the surface. Dye dilution and radiographic methods require the injection of certain materials into the blood stream. Flow measurement by heat transfer methods, involving a heat source, require major invasive procedures.

Ultrasonic blood-flow sensors employing the Doppler effect offer the advantage of completely noninvasive procedures. The measuring system consists of an oscillator, operating typically between 4 and 8 MHz, a pulsed power amplifier which drives a piezoelectric crystal at that frequency at periodic intervals, a piezoelectric crystal which senses the reflected RF energy, an amplifier for the received signal, and a means for detecting the difference between the transmitted and received frequency.

When the crystals are brought into close proximity to a blood vessel, at an angle θ to the centerline of the blood vessel, this frequency difference (the Doppler frequency, f_D) is proportional to the velocity (v) of blood flow through the vessel (due to the motion of red blood cells):

$$v = \frac{cf_D}{2f_t \cos \theta}$$

where c is the velocity of sound in blood and f_t is the frequency of the transmitted energy. The piezoelectric crystals perform the conversions of electrical-to-acoustic and acoustic-to-electrical energy. Separate crystals can be included in the (usually) hand-held probe, or a single crystal can be used alternatingly in the transmitting and receiving mode. The operating principle is illustrated in Figure 3-5 and a typical application of an ultrasonic blood-flow sensor is shown in Figure 3-6.

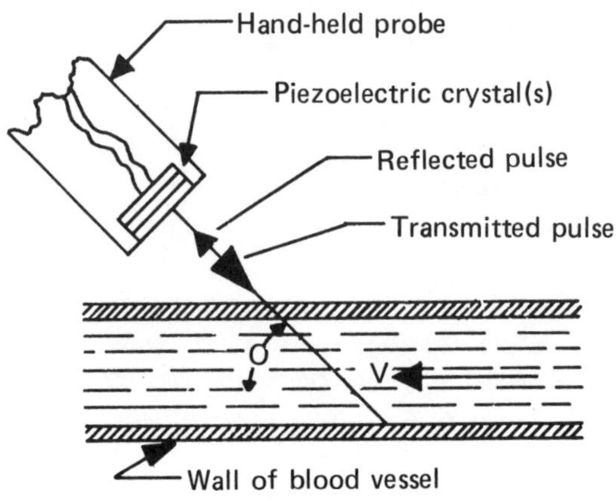

Figure 3-5. Ultrasonic (Doppler) blood flow measurement

A proper flow measurement taken with such a sensor requires a good zero-flow reference. If the volumetric flow rate (in liter or milliliters per second) must be determined instead of the flow velocity (in meters per second), the exact volume of the blood vessel must be known. Additionally, the direction of blood flow through the vessel must be known. If it is not known and needs to be determined, additional circuitry (two opposite-polarity phase shifters, each with its own mixer) can be used to indicate flow direction on the basis of two quadrature components of the transmitted RF signal. Typical pulse repetition rates are 10 to 30 kHz, with pulse durations in the low microsecond region, and maximum Doppler frequencies (at maximum flow rates up to about 4 m/s) are as high as 15 kHz. An aqueous jelly is used to provide good probe-to-skin contact.

The output signal of the sensing system (the Doppler frequency) is often fed to a loudspeaker for instantaneous, cursory monitoring. It is also fed to a frequency-to-dc converter whose output can be displayed (frequency meter) or recorded (strip-chart recorder).

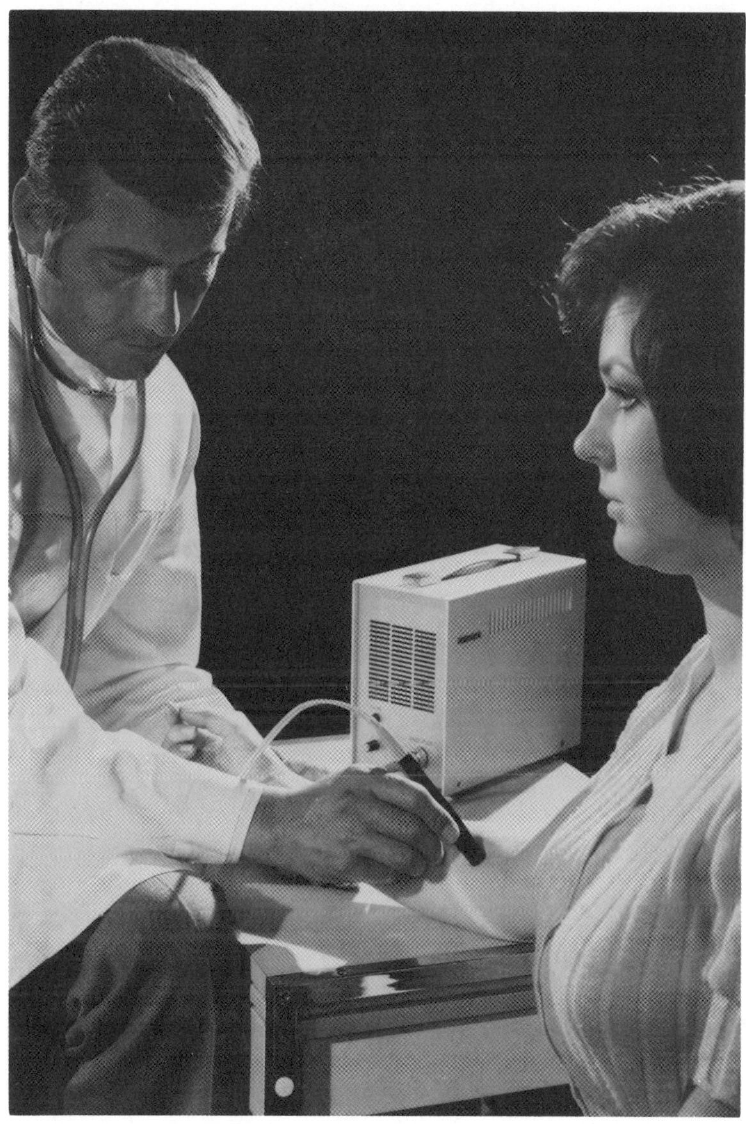

Figure 3-6. Application of ultrasonic blood flow sensor (courtesy of Siemens A.G.)

Ultrasonic blood-flow sensors employing two crystals at opposite wall surfaces across a blood vessel have been used at times when the need for an invasive procedure (to fully expose the blood vessel) could be tolerated. Ultrasonic sensors operating on the basis of transit time (shortened with increasing blood flow in the direction of energy transmission) rather than on the Doppler effect have also been designed but are not in common use.

Electromagnetic blood-flow sensors operate on the basis that blood is a conductive liquid. When such a fluid moves through a transverse magnetic field a voltage is induced in the fluid (the moving conductor). The magnetic field can be established by a permanent magnet or electromagnet around a blood vessel. The voltage is sensed by means of electrodes on the blood-vessel's external wall, at opposite sides of the wall and positioned normal to the magnetic field. Electromagnets with sinusoidal ac (400-1100 Hz) excitation are most commonly used.

Two types of electromagnetic blood flow sensors are in common use. Both incorporate the electromagnet as well as the two electrodes which provide the output signal. The *C-type* (or 'fish-hook') design (see Figure 3-7a) is slipped around a surgically-exposed blood vessel; the cover (black portion in the illustration) is then pushed into matching slots. This design is also implantable. The *cannula-type* design (cannulating transducer) requires that the blood vessel be severed and the two ends be attached over, and sutured to the two ports of the sensor (Figure 3-7b) in which the blood-vessel wall is simulated by a plastic conduit (cannula) penetrated by the electrodes. The *clip-on* design is a version of the C-type design, in this case supplied with a solid handle and mechanical means to actuate removal and replacement of the cover. The sensors are sized on the basis of their inside *(lumen)* diameter which can be between 1.5 and 40 mm. Sizing is somewhat less critical for *extracorporeal* blood-flow sensors which are similar in design to the cannula-type, but larger and heavier, and provided with metal ports to which plastic tubing can be attached.

The C-type sensors must be selected to fit tightly around the blood vessel. Most sensing systems include a set of sensors of different lumen diameters. A slight constriction of the blood vessel favors good electrode-to-vessel wall contact. A constriction of 10 to 20% is often recommended for acute studies, whereas a looser fit is favored for chronic implantations.

The C-type design is used for acute and chronic (implanted) applications. The clip-on design is intended for use during surgery. The cannula-type design is useable for acute and extracorporeal applications.

(a) C-type ("fish-hook") design, with slot cover in place

(b) Cannula-type design

Figure 3-7. Electromagnetic blood flow sensors (courtesy of Biotronex Laboratory, Inc.)

Figure 3-8 illustrates typical excitation/readout and ancillary equipment for electromagnetic blood-flow sensing systems. The excitation/readout unit provides the ac excitation for the electromagnet in the sensor, at a frequency selectable between 450 and 1050 Hz, and with a sinusoidal waveshape. The output signal (from the electrodes in the sensor) is accepted by the readout portion of the unit, fed through a gate and a low-pass filter, and displayed on the meter. The output signal differs in phase from the flow signal by 90°. The readout circuit includes provisions for an electronic zero-flow reference level (±5%). The gate, driven by the oscillator used for magnet excitation, permits detection of reversal of flow direction. Without the low-pass filter in the circuit, pulsatile flow is displayed. With the filter in the circuit, mean flow is displayed. The range indication can be selected as between 0-100 and 0-10 000 mℓ/min (0-10 ℓ/min) which is adequate for essentially all blood-flow measurements since normal aortic blood flow is 3.5-5 ℓ/min.

The analog computer shown in Figure 3-8 permits displays of derived quantities such as stroke volume ($\int Q_{dt}$), flow acceleration (dQ/dt) and additional quantities requiring the simultaneous output from a blood-pressure transducer (power and stroke work).

Thermal Blood-Flow Sensors–Thermal flowmeters (sensing transferred heat) have been applied to the measurement of blood flow rate (or velocity). Sensing systems incorporating an electric heater and a temperature sensor which detects the amount of cooling due to thermal convection caused by flow rate have been used experimentally. These require invasive procedures. A simpler and more promising method is given by heated resistance elements (anemometers), small enough to be placed at a catheter tip, whose resistance change due to cooling by blood flow is proportional to flow velocity. Such sensors can be of the type consisting of a thin resistive film on a subminiature glass substrate. A known amount of current heats the film and the sensor can be calibrated in terms of resistance (or voltage drop) vs flow rate causing the resistance change (decrease). Catheter-tip blood-flow anemometers require a substantially less invasive procedure than sensors relying on the insertion of a separate heat source in the blood vessel.

Blood-Flow Sensing with Injected Indicator Material–Blood flow (particularly in the aorta) has also been determined by noninvasive methods involving the injection of a detectable tracer material into the blood stream. Such materials include radioactive isotopes and colored dyes. The presence (and flow rate, while admixed to the blood) of radioactive isotopes can be sensed by means of radiation sensors. The colored dye, selected to have a colorimetric peak at a known wavelength, can be detected by means of optical-quantity sensors. Such indicator materials

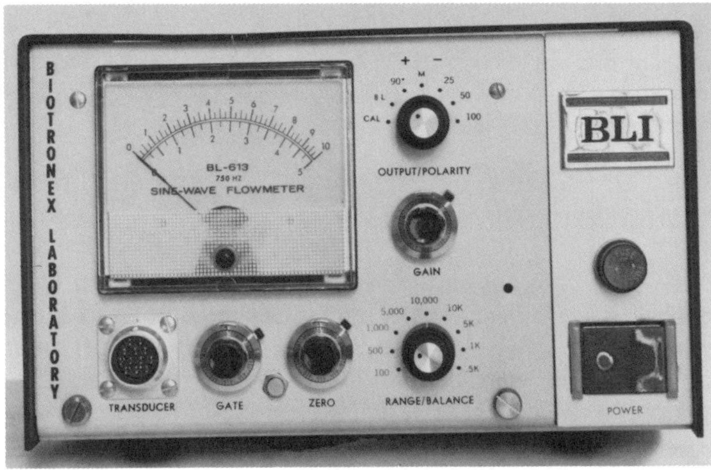

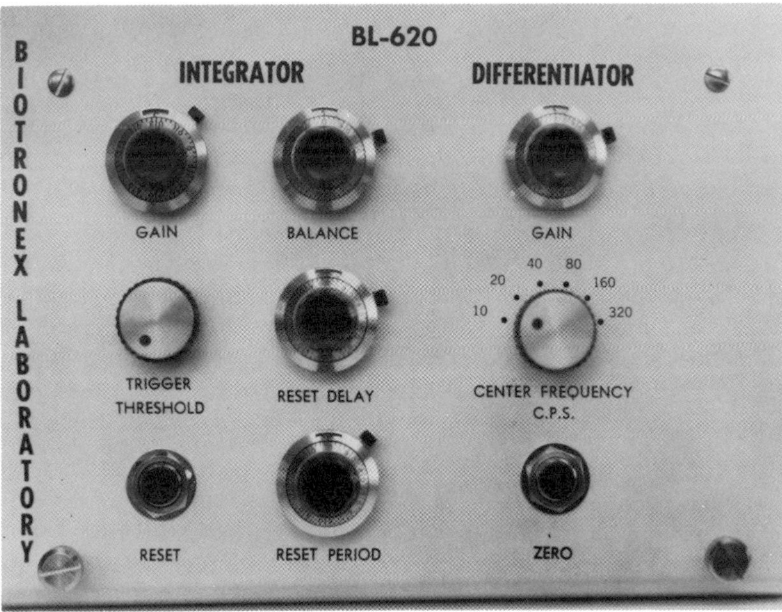

Figure 3-8. Excitation/readout unit (top) and inferred-quantity analog computer (bottom) for electromagnetic blood-flow sensing systems (courtesy of Biotronex Laboratories, Inc.)

must be selected on the basis of their nontoxicity and rapid elimination by the body systems. A safer method is *thermal dilution,* which involves the injection of an isotonic saline solution whose temperature is below body temperature, and detecting its concentration thermometrically (e.g., thermistor thermometer).

3.3 HEART POTENTIALS

Bioelectric action potentials originate at the sinoatrial (SA) node, are propagated multidirectionally along the surfaces of both atria, stimulate other action potentials when they reach the atrioventricular (AV) node, and cause those action potentials to travel through the walls of the ventricles with an appropriate time delay. The potentials effect contraction and relaxation of the atria and ventricles and the operation of the valves in the heart (see Figures 2-1 and 2-2).

The characteristics of these potentials, and their dynamic variations, can be determined by electrodes placed at various points of the body (arms, legs and chest) and connected to differential amplifiers whose output can be recorded on a strip-chart recorder or displayed on an oscilloscope. The recorder, with its amplifying, switching, and control circuits and its associated electrodes and interconnecting cables is called an *electrocardiograph* (ECG, or EKG on the basis of the Greek word *kardia,* heart). The graphic display is an *electrocardiogram.* Electrocardiographs (recorder-type) are widely used for the diagnosis of anomalies in heart action, such as arrhythmia and myocardial damage.

Figure 3-9 illustrates a typical cycle of an electrocardiogram. Although there are a number of acceptable tolerances on amplitude and time of waves and intervals (with amplitudes partly determined by coupling of signals through surface electrodes), the dynamic variations and relative slopes of the waves, in addition to major deviations in relative amplitude and time intervals, can be used for definitive diagnoses of heart-action anomalies or, more generally, for determination of the state of health of the heart.

Since the heart is a three-dimensional organ which moves constantly along each of three axes, heart-action potentials are sensed by electrodes placed at various points of the body as an approach to obtaining a vectorial representation. Routinely established and now virtually standardized electrode placement points and their relative polarities and connection to an electrocardiograph are illustrated in Figure 3-10.

The basic *(bipolar)* limb leads (I, II and III) are based on the *Einthoven Triangle* formed by the shoulders and the crotch, with the heart at the

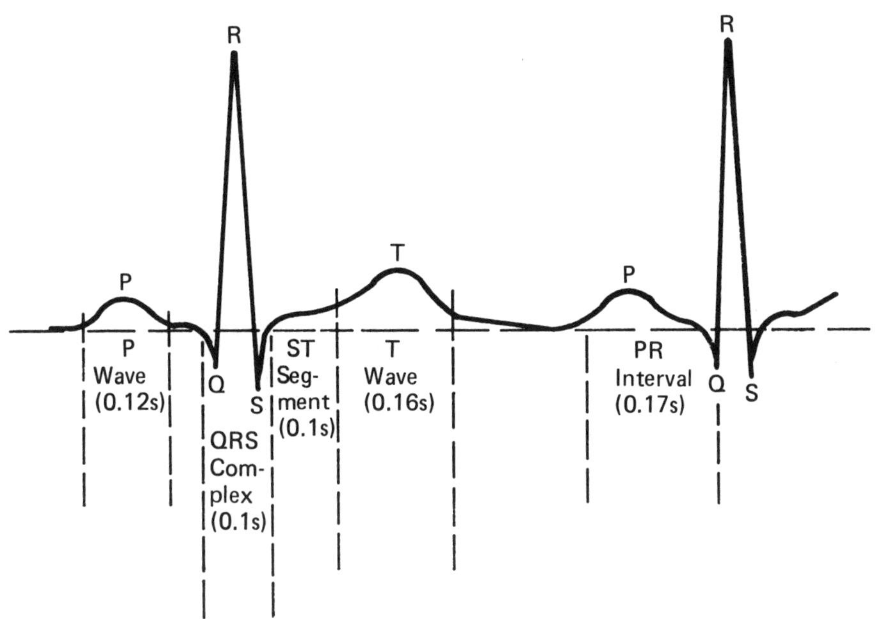

P Wave - depolarization (excitation) of atria
PR Interval - impulse propagates through atria and AV node
QRS Complex - repolarization of atria and depolarization (excitation) of ventricles
ST Segment - period between ventricular depolarization and start of repolarization
T Wave - ventricular repolarization (recovery)

Note: Time intervals are typical or average values

Figure 3-9. Electrocardiographic cycle

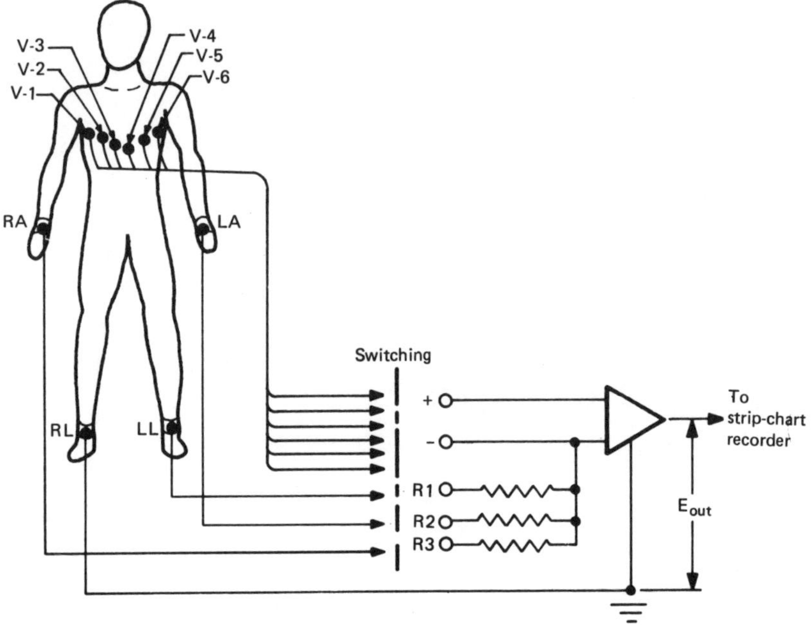

EKG Channel (Switch Position)	Electrode-to-Terminal Connection								
	RA	LA	LL	V-1	V-2	V-3	V-4	V-5	V-6
I (or L-I)	-	+							
II (or L-II)	-		+						
III (or L-III)		-	+						
aVR (or AVR)	+	R2	R3						
aVL (or AVL)	R1	+	R3						
aVF (or AVF)	R1	R2	+						
V_1 (or V-1)	R1	R2	R3	+					
V_2 (or V-2)	R1	R2	R3		+				
V_3 (or V-3)	R1	R2	R3			+			
V_4 (or V-4)	R1	R2	R3				+		
V_5 (or V-5)	R1	R2	R3					+	
V_6 (or V-6)	R1	R2	R3						+

Figure 3-10. EKG electrode placement and lead connections

approximate center of this roughly equilateral triangle. Since potentials at the shoulders are nearly the same as at the wrists, and potentials at the ankles close to those at the crotch, for EKG purposes, the electrodes are placed at those more conveniently accessible points. *Unipolar* measurements can be obtained by connecting resistors (about 5 kΩ) in the negative limb leads so that the positive electrode is referenced to the average of the potentials sensed by two or three other electrodes.

In the case of the *augmented* limb leads (aVR, aVL, aVF) one limb electrode is referenced to the average of the potentials at the other two limbs. In the case of *chest* (precordial) *leads* (V_1 through V_6) the chest electrode is referenced to the average of the three limb potentials. The EKG input circuit is a differential amplifier which must have a very good common-mode rejection.

Electrocardiographs provide the lead-selection and resistor switching, signal amplification with a frequency response between quasi-dc (about 0.05 Hz) to 100 Hz (the upper-frequency cutoff is sometimes selectable), a high input impedance (10 megohms or greater), good isolation of any components in contact with the patient from chassis ground and related patient-safety provisions, and a strip-chart recorder (one, three or six channels) with pen-position and gain adjustments, chart-speed selection, calibration signals and, sometimes, built-in trace (lead) identification. Two models are shown in Figure 3-11. A number of portable, battery-operated designs are available from a variety of manufacturers in many parts of the world. Many of these have an ac adapter and battery recharger. Other models are intended for line-voltage operation, sometimes while part of a console also containing other instruments.

The electrodes are attached to the patient before the EKG is operated. Typical designs of the two most commonly used electrodes, the plate (limb) electrode and the suction (vacuum) electrode, are shown in Figure 3-12. The portion of the skin which will be in contact with an electrode is first coated with an electrolyte jelly to provide an interface between the ionic potentials and currents in the body and the input of the amplifier. Plate electrodes are attached with rubber or canvas straps. Suction electrodes are attached by first squeezing the rubber bulb, then placing the electrode firmly against the skin, finally releasing the bulb to obtain a partial vacuum between skin and electrode.

After all electrodes are attached, the connecting leads from the EKG are clipped to the contacts on the appropriate electrodes, the EKG is checked and calibrated, and recordings are obtained with sequential selection of lead position (channel). Only a few seconds are required to obtain a recording for each lead position. The resulting traces are identi-

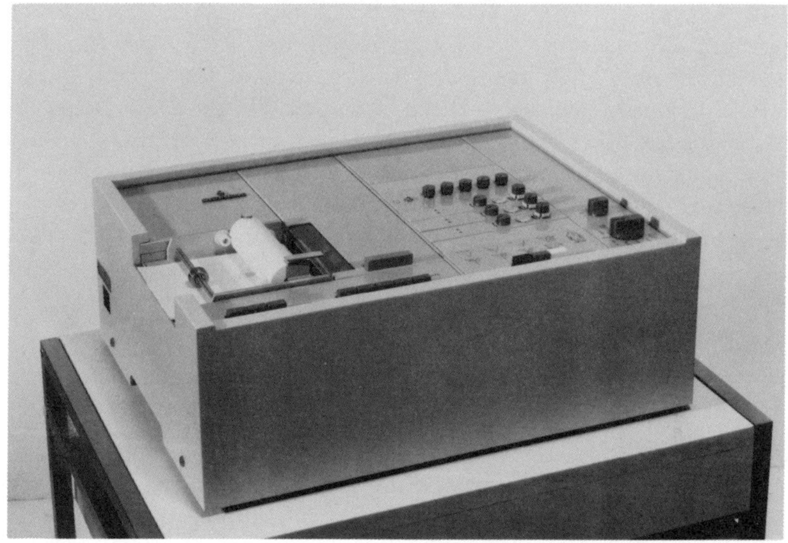

(a) Three-channel electrocardiograph

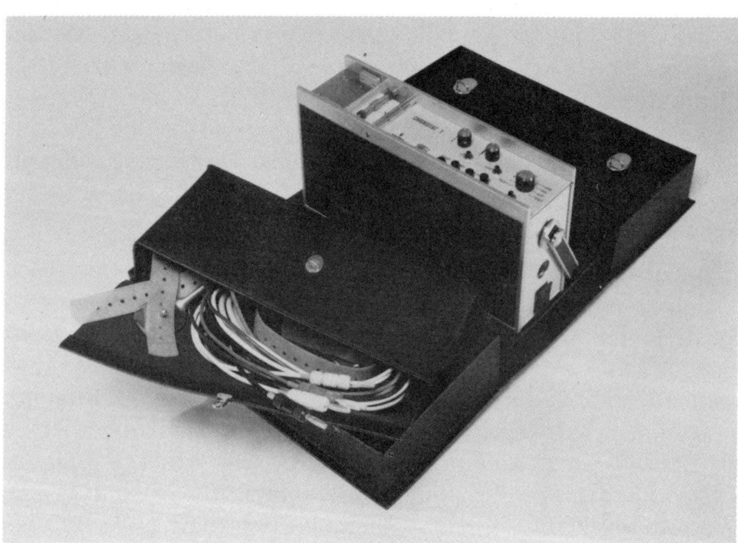

(b) Single-channel electrocardiograph with accessories (straps, electrodes, leads) and carrying case

Figure 3-11. Typical electrocardiographs (courtesy of Siemens A.G.)

(a) Plate electrode

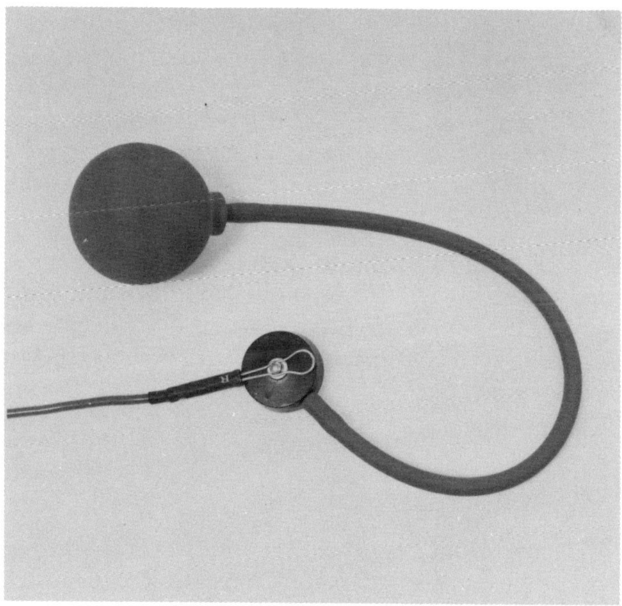

(b) Suction electrode

Figure 3-12. EKG electrodes (courtesy of Siemens A.G.)

fied, the strip chart is cut and usually mounted in a standard format: I, II, III; aVR, aVL, aVF; then (below) V1, V2, V3; and V4, V5, V6. Use of a three-channel electrocardiograph facilitates this form of display.

Heart rate can be easily derived from the EKG output (number of electrocardiographic cycles per unit time). Ancillary equipment can be used to display derived heart rate by means of an analog or digital readout device *(cardiotachometer),* or by causing a lamp to flash or an audible tone to be heard for each heartbeat. The latter scheme is often used in hospital coronary care units and operating rooms. Another electrocardiographically derived display is the *vectorcardiogram,* an approximate presentation of the locus of the cardiac vector obtained by using one lead output to drive the X-axis and another lead output to drive the Y-axis of an oscilloscope, each output so chosen as to represent a cardiac vectorial component.

Automatic EKG evaluation, by use of a computer, has been developed, and implemented at many locations, with the intent of assisting physicians with routine analyses of electrocardiograms (it is not likely that the diagnostic capabilities of an experienced physician will ever be replaced by a computer). EKG-evaluation computer programs are usually based on inputs from the standard 12-lead configuration and additional inputs such as height, weight, sex, blood pressure and any drug-taking habits of the patient. These inputs are first recorded on magnetic tape, to permit further processing as well as to allow data recall. The EKG signals are then digitized and the digital data are evaluated by special computer programs. The output is a tabular numerical print-out of the signal characteristics and their interrelationships, as well as a language-text evaluation of the results, especially of any anomalies detected.

Large facilities, such as hospitals or clinics, may employ a computer wholly or partly dedicated to EKG evaluation. Smaller facilities, including doctors' offices can obtain access to a remotely-located computer by means of a phone terminal incorporated into the electrocardiograph as ancillary equipment.

3.4 HEART SOUNDS

Heart sounds are, primarily, hydrodynamically-caused vibrations in that portion of the cardiovascular system consisting of the heart and the immediately adjoining major veins and arteries. The term *auscultation* (a Latin-derived word meaning "listening") is applied to the detection of sounds created by the motions of organs of the human body, including the heart. A stethoscope is commonly employed for auscultation. However, most of the cardiac vibratory spectrum is below the threshold of audibility. Also, the sounds tend to be modified within the stethoscope and the human ear is a less-than-perfect receiver of such sounds. *Phonocardiography,* the electronic sensing, amplification and (graphic) display of heart sounds, over a wide frequency spectrum, is used instead of stethoscopic auscultation when a more detailed and accurate knowledge of heart sounds is required.

The time relationship of heart sounds to blood pressure, valve events and heart potentials is illustrated in Figure 3-13. The first two heart sounds ("lubb-dub") have sufficient audible components and duration to be easily detectable by non-electronic means. The *first sound* is the result of ventricular contraction, closure of the atrioventricular (tricuspid and mitral) valves, opening of the semilunar (pulmonary and aortic) valves, and blood ejection. The frequency spectrum of the first sound is from 30 to 150 Hz and the duration of the sound is 100 to 160 ms. The *second sound,* with a duration of 80-140 ms and a frequency range of 225-400 Hz, is the result of the closure of the pulmonary and aortic valves. The *third sound,* usually heard only after exercise, has a frequency range of 10-100 Hz and a duration of 40-80 ms, and is ascribed to ventricular-wall vibration resulting from the inrush of blood into the ventricles. A fourth sound (not shown in Figure 3-13) is essentially inaudible but phonocardiographically detectable. It is the result of atrial contraction *(atrial heart sound)* and is characterized by a frequency range of 10-50 Hz and a duration of 30-60 ms.

Phonocardiography equipment, consisting of a microphone with cable (see Figure 3-14) and a preamplifier/conditioner unit, is commonly available as equipment ancillary to electrocardiographs which, when of the multichannel type, provide the final amplification and display capability. Microphones are predominantly of the piezoelectric type and can be air-coupled (small fixed volume of air between skin and diaphragm) or directly contacting (mechanical contact between diaphragm and skin). The preamplifier usually contains switch-selectable low-frequency-suppression filters. These facilitate the analysis of higher-frequency components of importance in detecting anomalies such as heart murmurs. The transducer frequency response is typically from less than 2 to over 1000 Hz.

Phonocardiography has been found particularly useful for the detection of such heart anomalies as *stenosis* (narrowing of the mitral or aortic valve opening), *regurgitation* (incomplete valve closure resulting in partial backward flow of blood) and other heart-valve insufficiencies. These can be diagnosed with the aid of phonocardiograms containing

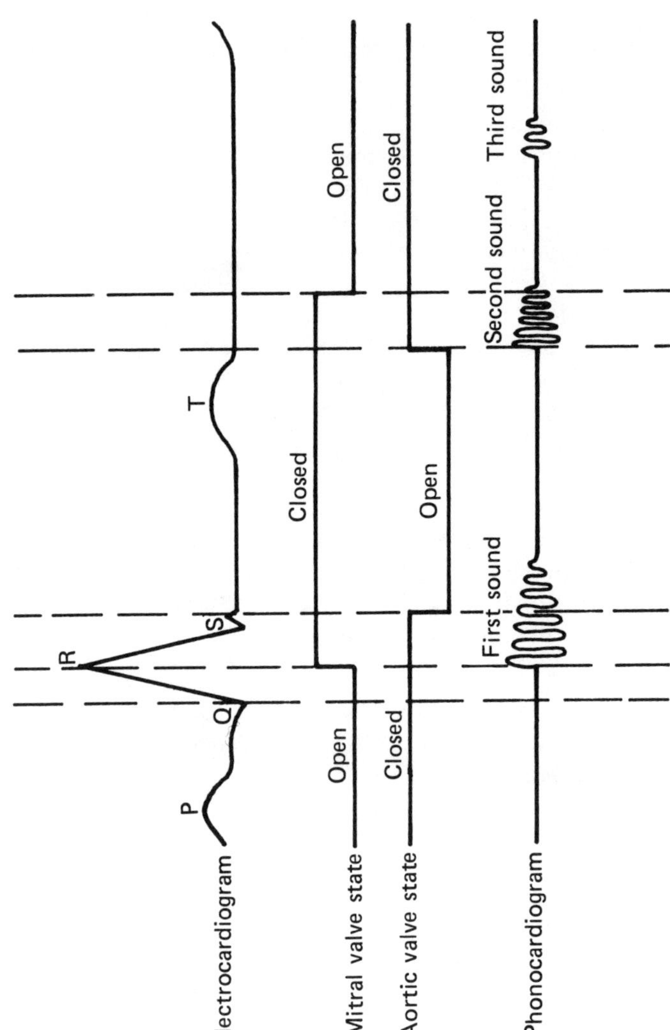

Figure 3-13. Time relationship of heart sounds to EKG and heart valve status

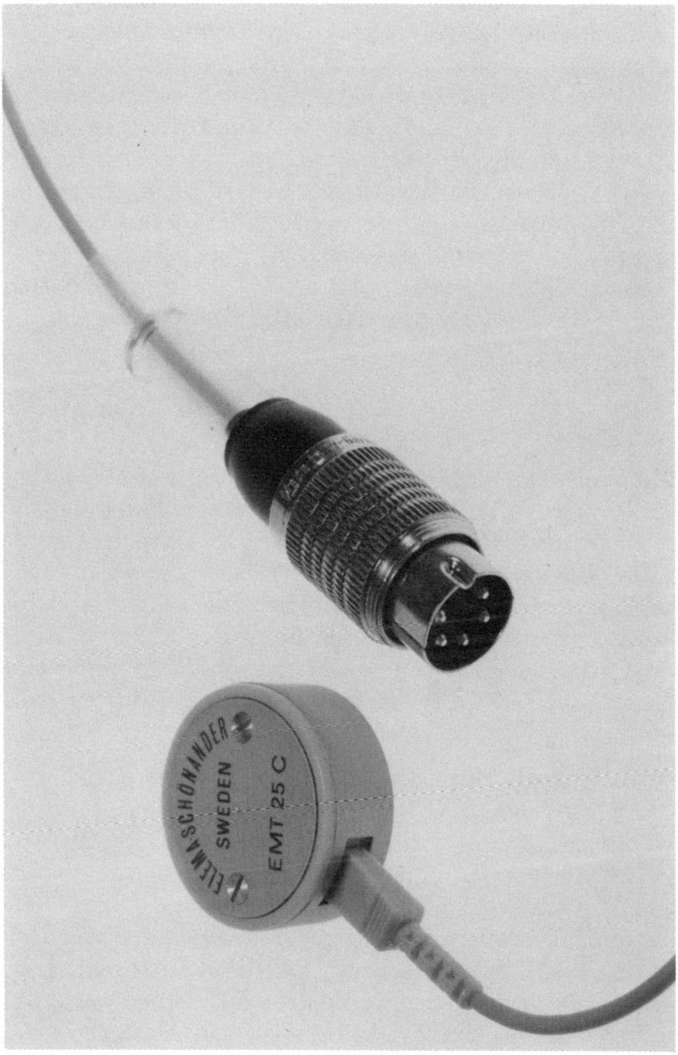

Figure 3-14. Phonocardiography microphone (courtesy of Siemens-Elema A/B)

such abnormal sounds as murmurs, gallops, clicks (systolic) and snaps (diastolic) characterized by frequency components in the 50-100 Hz region.

Intercardiac phonocardiography, infrequency used in other than research procedures, requires the use of a catheter-tip microphone, several types of which have been developed. *Spectral phonocardiograms* require somewhat different signal conditioning to facilitate the direct determination of sound-pressure level (in dB) and frequency composition of heart sounds. They are useful in checking the operation of cardiac valve prostheses and in detecting myocardial dysfunction. A *vibrocardiogram* is the display of vibrations at the thoracic surface, resulting from the impact of the heart against the thoracic cavity. These vibrations are characterized by a predominance of low-frequency components. Vibrations caused by the apex of the heart impacting on the rib cage, sensed close to the location of the apex, are displayed as an *apex cardiogram.*

3.5 PULSE

The sequential expansion and contraction of the walls of blood vessels, the pulse, is a major diagnostic indicator. For simple diagnostic procedures, such as doctors' offices, it is customary to determine pulse rate by feeling the pulse of the radial artery (above the wrist) and counting the pulse for 30 seconds, then multiplying the count by two, to obtain pulse rate per minute. More detailed pulse determination, including waveshape and frequency-component analysis and amplitude measurements, can be performed by means of electronic sensing and display systems.

Pulse sensors are, typically, displacement transducers with a medium-high frequency response (e.g., strain-gage or reluctive) whose sensing element is in mechanical contact with the surface of the skin immediately above the artery or vein being sensed. The sensor is fastened to the body by means of a strap or tape, or it is mounted to a stand, then adjusted until its sensing element is in firm contact with the point of measurement. Figure 3-15a shows a sensor for arterial pulse measurements. The sensing element of a venous pulse sensor, which much be capable of responding to much smaller displacements than arterial sensors, is shown in Figure 3-15b. A photoelectric (light beam-optics-light sensor) technique is used for displacement sensing in the pulse sensor shown in Figure 3-15c which is used while mounted on a stand, then adjusted until the gap between it and the point of measurement is about 2 mm. Piezoelectric pulse sensors have a low-frequency cutoff at about 2 Hz but have been used for peripheral pulse measurements.

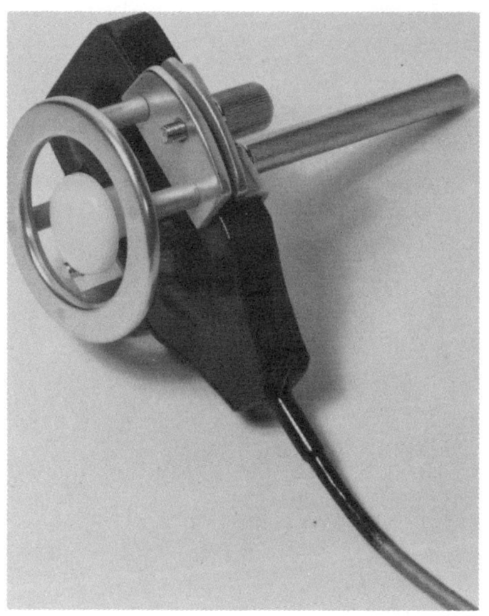

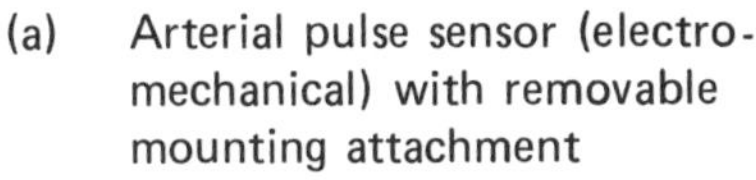

(a) Arterial pulse sensor (electro-mechanical) with removable mounting attachment

(b) Sensing element of venous pulse sensor

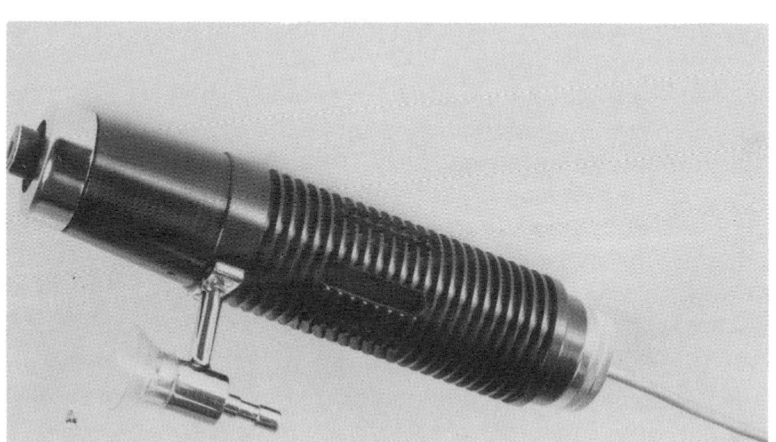

(c) Photoelectric pulse sensor

Figure 3-15. Pulse sensors (courtesy of Siemens A.G.)

Arterial pulse measurements are typically taken from the carotid or the subclavian artery, also from various parts of the extremities, such as the brachial, radial, femoral and popliteal artery. The venous pulse is mostly taken at the right or left external jugular vein. The pulse is also measured on fingers and toes, using sensors of slightly different configuration. Figure 3-16 shows a typical installation of an arterial pulse sensor.

Pulse measurements obtained with sensors such as those described above are usually displayed on a strip-chart recorder. Figure 3-17a illustrates a normal arterial pulse recording. A suitably amplified venous pulse recording is shown in Figure 3-17b, correlatable to an EKG and heart sounds of the same (healthy) subject. The outputs of two or more pulse sensors, proximal and distal to the heart, can be recorded simultaneously for determinations of the travel of the pulse wave.

The pulse rate can be determined by employing ancillary electronic circuitry which extracts the time between two consecutive pulse waves and converts this information into a display of beats per minute. Such circuitry can also obtain its input from EKG eqipment. A device which displays heart rate (heart beats per minute) is called a *cardiotachometer*. Methods other than the one described here have also been used to display pulse rate, e.g., integrating a number of pulse-wave peaks over a fixed period of time.

3.6 OTHER CARDIOVASCULAR-SYSTEM MEASUREMENTS

Plethysmography is the measurement (and display) of variations in the size of an organ, an extremity or another part of the body, which are due to cardiovascular action (blood volume changes). The simplest plethysmographs and nonelastic cups sealed to a finger or toe, or a chamber sealed to a limb (e.g., under the elbow or under the knee). The chamber or cup can be filled with either liquid or air and is ported to a pressure transducer which senses the volume fluctuations in terms of variations of pressure. Alternatively, the volume changes can be sensed in terms of displacement.

A *capacitance plethysmograph* measures volume changes in terms of changes in capacitance between the skin surface of the part to be measured and a metallic sleeve insulated from the skin by a dielectric. The skin is one electrode of a capacitor, the sleeve is the other electrode, and the variations in capacitance are converted into voltage changes which can be displayed.

A *photoelectric plethysmograph* uses a sensor placed on the skin di-

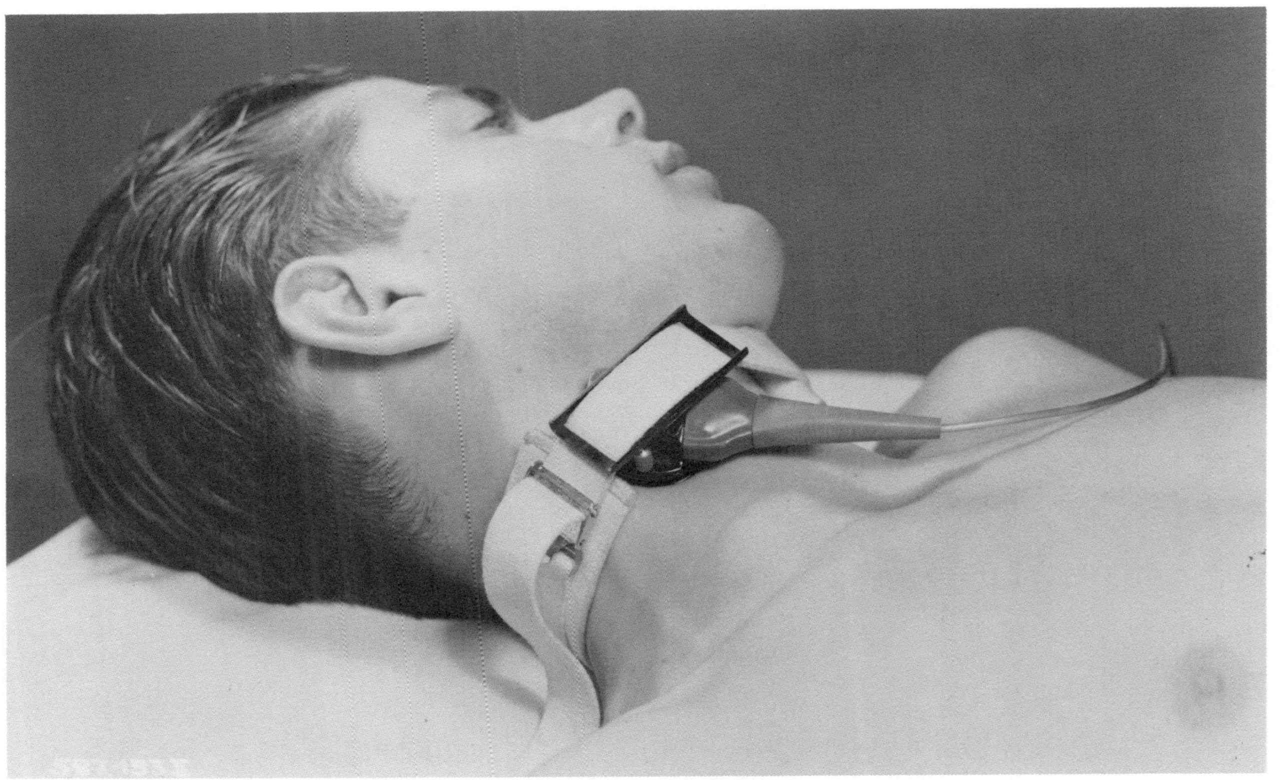

Figure 3-16. Pulse sensor strapped in place on carotid artery (courtesy of Siemens A.G.)

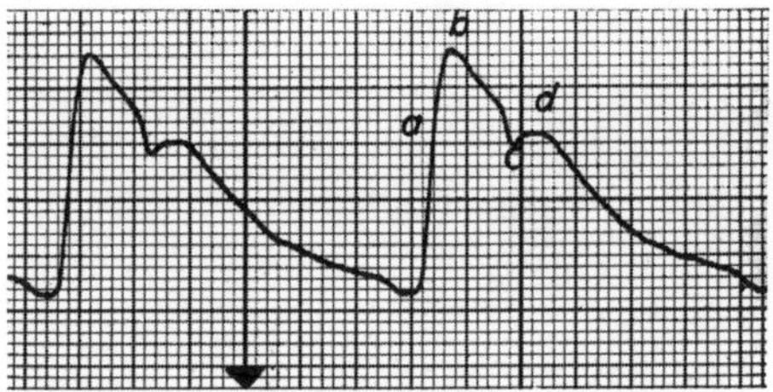

(a) Arterial pulse, taken from the large carotid artery

a - systolic rise
b - initial oscillation
c - cleft
d - residual oscillation

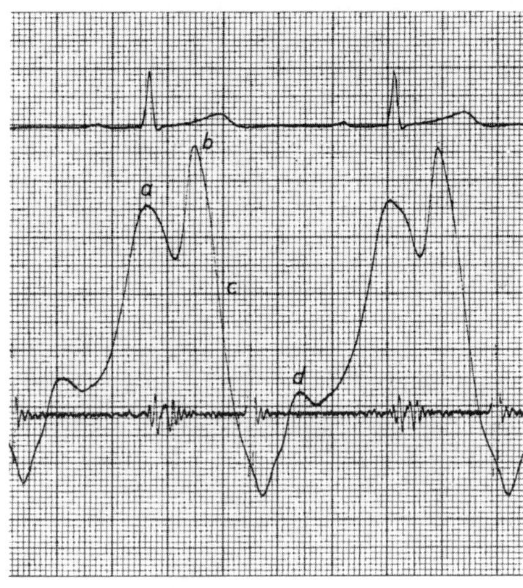

(b) Venous pulse, synchronized with EKG (top) and heart sounds (bottom)

a - presystolic wave
b - systolic wave
c - systolic collapse
d - diastolic wave

Figure 3-17. Typical pulse recordings for healthy cardiovascular system (courtesy of Siemens A.G.)

rectly over a blood vessel and measures blood volume fluctuations in terms of opacity variations. Typical points of application are finger tips and the external ear.

The *impedance plethysmograph* measures volume changes in terms of the change of bulk impedance between electrodes placed at two or more points of the skin surface of the part to be measured. Typically, a very small current from a constant-current source is passed through the measured part by means of the electrodes and the impedance variations are sensed in terms of voltage variations. Microwave systems have been used for impedance plethysmography, using transmitting and receiving apertures in lieu of electrodes.

Ultrasonic techniques are frequently used to determine characteristics of the heart and circulatory system. An ultrasonic transceiver is typically placed over the part to be measured and pulsed ultrasonic energy is directed at selected portions of the part. Ultrasonic imaging displays of the part can be obtained by measuring the time between transmission of a pulse and receipt of its echo. Motion of portions of the part can be detected by Doppler techniques or by a scanning technique using pulsed ultrasound with an M-scan in the readout.

Thus, the *echocardiogram* provides a noninvasive technique for determination of heart valve functions and heart cavity displacements. One of its specific applications is the evaluation of the performance of cardiac valve prostheses. *Echopericardiograms* can be used to differentiate pericardial effusion from cardiomegaly. An *echoaortagram* can detect aortic aneurysms and show the aortic wall thickness. Doppler techniques *(doppler sonograms)* have been found particularly useful in recording foetal heart beat. They have also helped in isolating obstructions in peripheral circulation.

Figure 3-18 illustrates the application of ultrasonic Doppler techniques to the detection of foetal heart rate. This equipment employs, additionally, a sensor for the (simultaneous) measurement of uterine pressure during labor.

Ultrasonic techniques have largely replaced X-ray methods for displays of heart action, such as used in *cine X-ray heart profile recorders* which provide motion pictures of the heart using X-rays as illumination.

Ballistocardiograms provide an indirect display of heart action. They are obtained by placing the subject on a platform which is restrained from motion only by transducer-equipped linkages so that the dynamic responses of the body to the beating and blood-pumping heart can be sensed in terms of force or displacement.

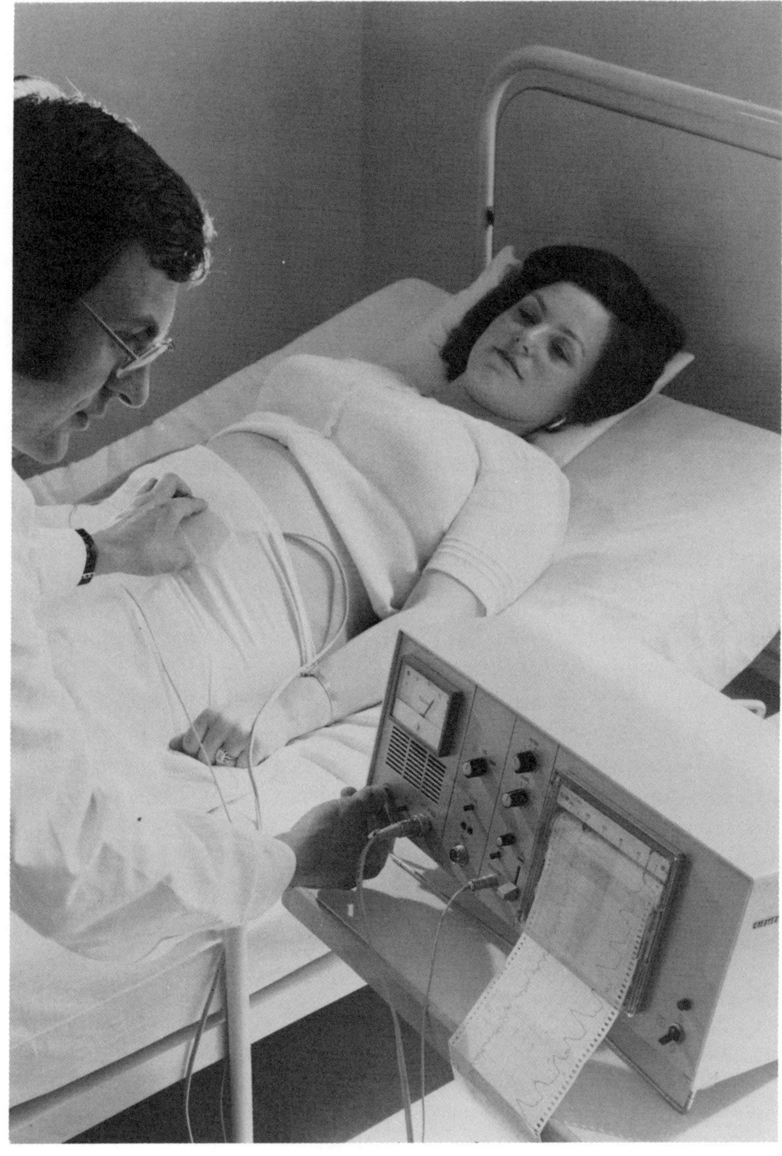

Figure 3-18. Ultrasonic Doppler sonogram equipment for foetal heart rate detection (courtesy of Siemens A.G.)

4. Neuromuscular System Measurements

Bioelectric potentials in the nervous and muscular system (see 2.3) are sensed by means of *electrodes* which are connected to electronic amplification, signal conditioning and display equipment. Electrodes are also used to apply low-voltage, low-current stimuli to nerves and muscles so that specific responses to such stimuli can be obtained.

4.1 ELECTRODES

The function of a (sensing) electrode is to convert ionic potentials into electronic potentials. The signal obtained is always the difference between the potentials at two electrodes. In some measuring set-ups one electrode is used as reference electrode and the potentials between it and each of several other electrodes are observed.

Three types of electrodes are available:

(a) *skin (surface) electrodes,* usually in the shape of small (1 cm diameter or less) discs to which a lead or terminal is permanently affixed;

(b) *needle electrodes,* as illustrated in Figure 4-1, usually less than 1 mm diameter, with a sharp point;

(c) *microelectrodes,* a special form of needle electrodes, with tip diameters as small as one μm.

Electrodes used primarily for EKG are covered in 3.3.

The most commonly used electrode material is silver, covered with

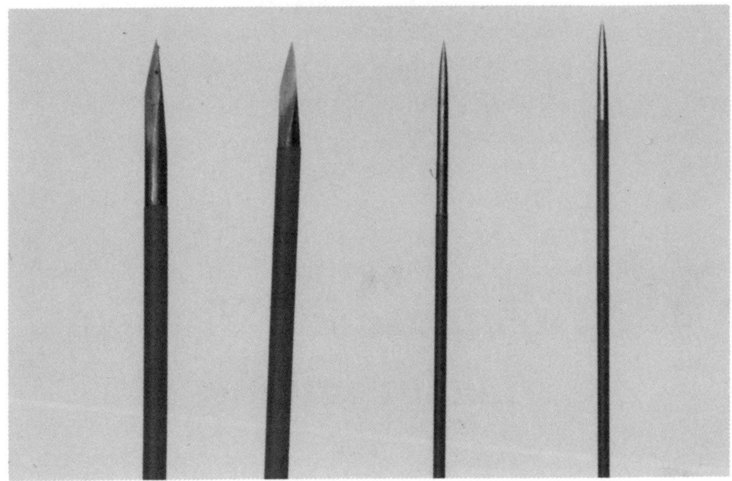

(a) Typical Teflon-coated electrodes

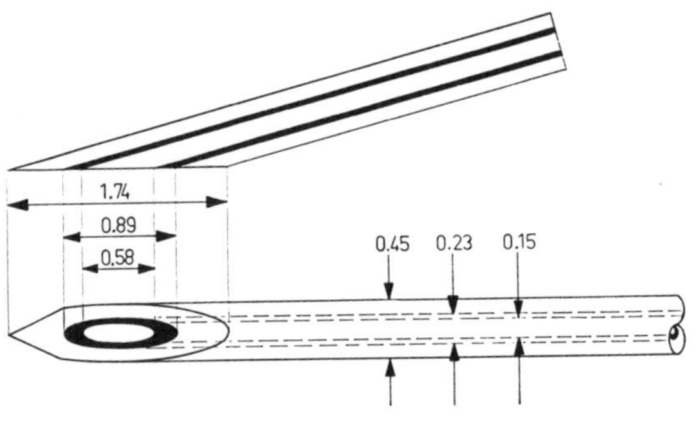

(b) Concentric electrode

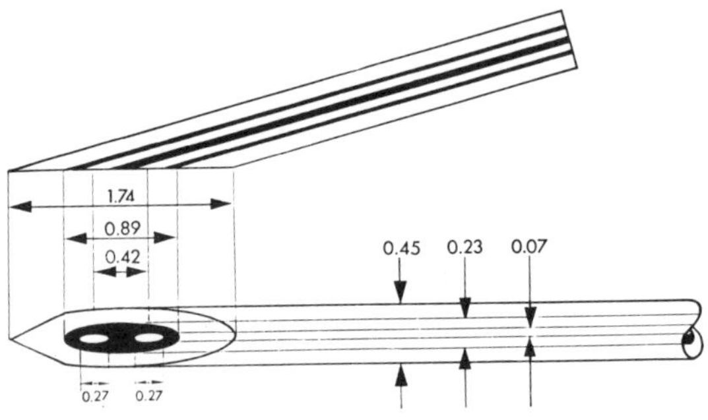

(c) Bipolar electrode

Figure 4-1. Needle electrodes (courtesy of DISA Elektronik A/S)

silver chloride. Platinum is often used for muscle potentials, which are significantly larger than the typical nerve potentials. An electrolyte paste or jelly is usually introduced between the point of measurement and the electrode.

A special version of the skin electrode, the *floating* type, is a small inverted cup filled with electrolyte jelly and so constructed that the electrolyte is the sole contact material. A *multiple electrode* is a needle electrode with segments of the needle insulated from each other and each provided with a separate lead, so that simultaneous measurements can be obtained at several depths below the skin. A *pharyngeal electrode* is a thin (about 1.5 mm dia.) insulated silver wire ending in a noninsulated small silver ball; this electrode is intended for insertion through the nasal passage.

Most electrodes are unipolar and require the use of a separate reference electrode. Exceptions are the concentric and bipolar electrodes (Figure 4-1). The *concentric electrode* contains a central conductor insulated from its sleeve so that the sleeve can act as reference electrode. The *bipolar electrode* contains two conductors, insulated from each other as well as from the sleeve; it is available with varying amounts of spacing between conductors so that potentials at two closely-spaced points of measurement can be obtained with a known amount of spacing between the points. The signal can then be the difference between the two potentials.

Microelectrodes have been developed in two design versions: solid (e.g., metal on glass) or liquid-filled. The latter are in the form of very thin hollow glass tubes filled with a solution usually containing chloride ions. It should be noted that the small tip size of a microelectrode makes it suitable for recording the activity of a single nerve cell.

Since nerve and muscle potentials are very low (generally less than 200 μV for nerve potentials, less than 5 mV for muscle potentials) proper precautions must always be taken to minimize noise pickup at the electrode and to recognize and avoid (or electrically compensate for) spurious voltages generated at the electrode, including erroneous potentials due to polarization which results from dc current flow through the metal/electrolyte interface.

Because the impedance between two electrodes is very high (in the tens of kilohms range for skin electrodes and in the tens, hundreds or thousands of kilohms for needle and microelectrodes), it is essential that the amplifier used with the electrodes have a suitable high input impedance and good common-mode rejection as well as the proper frequency response.

4.2 BRAIN POTENTIALS

The electrical activity of the brain can be observed from measurements of neurogenic brain potentials on the scalp. This measurement technique is called *electroencephalography* (from the Greek word *kephale,* head) and the resulting recording of the signals is an *electroencephalogram* (EEG). The measuring and display system is an *electroencephalograph* (EEG).

Either skin surface electrodes, sometimes of the floating type and usually of silver/silver chloride, or needle electrodes are used to obtain potentials at various points on the scalp. It should be noted that the needles do not penetrate the skull, but extend only a very short distance below the skin surface. The measured potentials are actually not action potentials of the brain nerve cells, but are representative of synchronized graded potentials of neurons in the vicinity of the electrodes.

The electrodes are normally placed on the scalp in accordance with a standard placement pattern called the *10-20 system* because this spacing is at intervals of 10% and 20% of the distance between specified points on the scalp. Each location is designated by a letter and number. The letter refers to the lobe of the brain (e.g., F = frontal, T = temporal, 0 = occipital) and the number to a standardized location over that lobe. Thus, "T3" is a location on the temporal lobe just above the left ear. This system is reflected in the pin designations of an electroencephalograph input adapter (Figure 4-2) to which thin, flexible, insulated leads from each of the electrodes are connected. This adapter can be attached to a rigid arm or stand and is furnished with a multiconductor cable (shown in the illustration) with a connector which plugs into the electroencephalograph.

EEG potentials are generally characterized by amplitudes from about 5 to 200 μV and frequencies from about 0.1 to 100 Hz. Typical recordings show random-appearing waveforms but four waveforms in different frequency bands have been identified. The most prominent of these is the *alpha* wave (7.5-13 Hz, 10-100 μV). It should be noted that this wave is more pronounced when the patient is at rest, with eyes closed and mind wandering freely. The *theta* wave, usually well pronounced in EEGs of normal children, has a frequency band of 3.5-7.5 Hz and amplitude variations of 50-200 μV. Two additional waves of lower amplitude (10-50 μV) have been identified as the *beta* wave (13-30 Hz) and the infrequently observed *delta* wave (0.2-3.5 Hz).

The nature of the EEG signals imposes requirements on electroencephalographs not only for preamplifiers with appropriate gain and frequency response, as well as a high input impedance, but also for sensi-

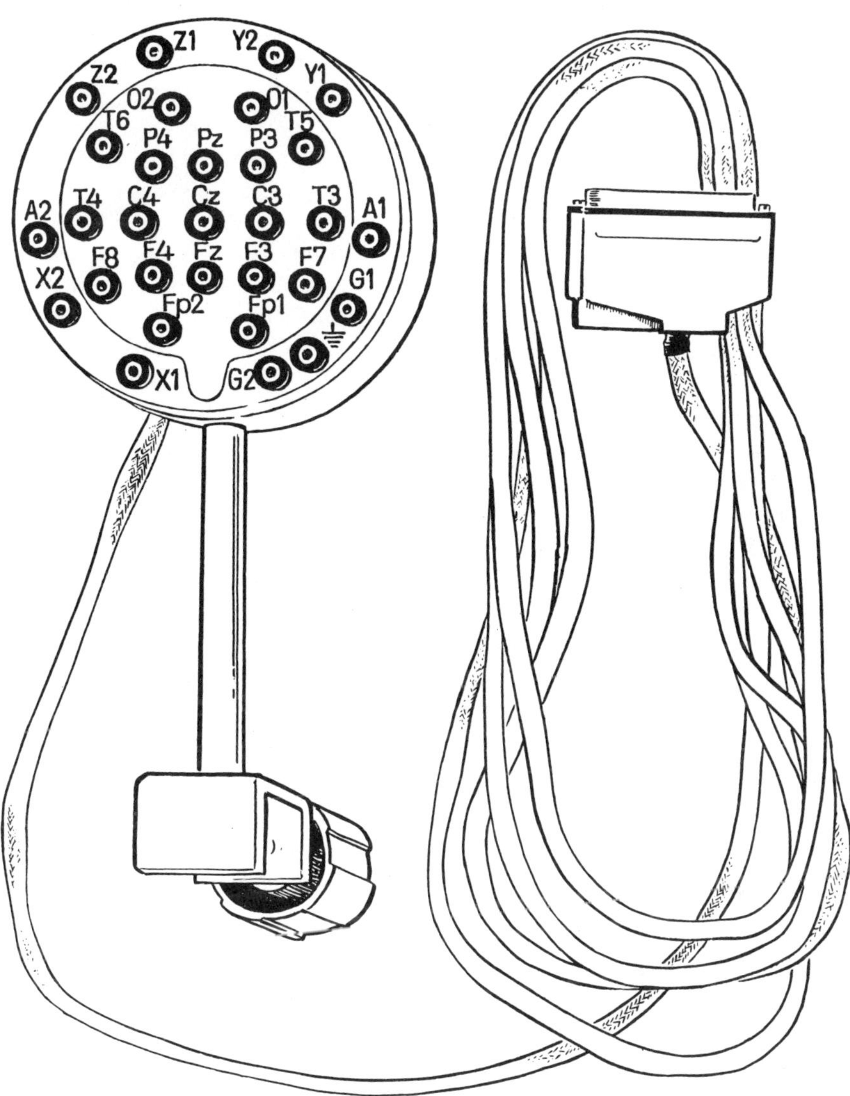

Figure 4-2. Electroencephalograph input adapter (courtesy of Siemens-Elema AB)

tivity and frequency-response selectability so that data reduction is facilitated. It is desirable that potentials from several electrodes can be displayed simultaneously. 8-channel EEGs are often used, but EEGs with up to 16 channels (Figure 4-3) are available. Good shielding and grounding practices are imperative for EEG operation. A shielded room can be used, at least for the patient and the input wiring, to reduce interference effects.

4.3 MUSCLE POTENTIALS

The monitoring of the electrical activity of muscles *(electromyography)* involves the sensing, conditioning and display of potentials generated in muscle fibers when they contract in response to a stimulus. The display of muscle potentials is called an *electromyogram* (EMG). The potentials are averaged action potentials of groups of muscle cells. The stimulus can either emanate from the subject's nervous system *(voluntary-effort potentials)* or from an externally applied electrical-current pulse *(evoked potentials)*. EMGs are used for diagnoses of nervous and muscular disorders.

Externally-applied stimuli are applied to the skin immediately above the muscle by means of a stimulation electrode (Figure 4-4a). Figure 4-4b shows a typical stimulus control unit which permits adjustments of stimulus amplitude, pulse width, repetition rate, and duration of a series of repeated stimuli (pulse trains). Either pulse trains or single pulses can be applied.

Electrodes, of either the needle or surface type, are used for sensing muscle potentials. Needle electrodes permit better localization of the muscle area to be sensed. Surface electrodes are generally used only for sensing large-muscle activity. Good skin contact is essential for both types of electrodes.

The signal conditioning and recording equipment, typically using an ungrounded bipolar-lead feeding into a differential amplifier, should provide a frequency-response capability of zero to over 1000 Hz and be capable of handling the approximately 0.02 to 5 mV amplitude range of EMG signals.

Figure 4-5 illustrates an electromyograph which includes a stimulus control unit. The equipment shown provides for volatile as well as nonvolatile display of up to four EMG signals (the Digigraph is used for permanent recordings). The Digital Control Unit allows storage of up to four EMG signals in its memory. Recordings can thus be made of real-time or stored signals.

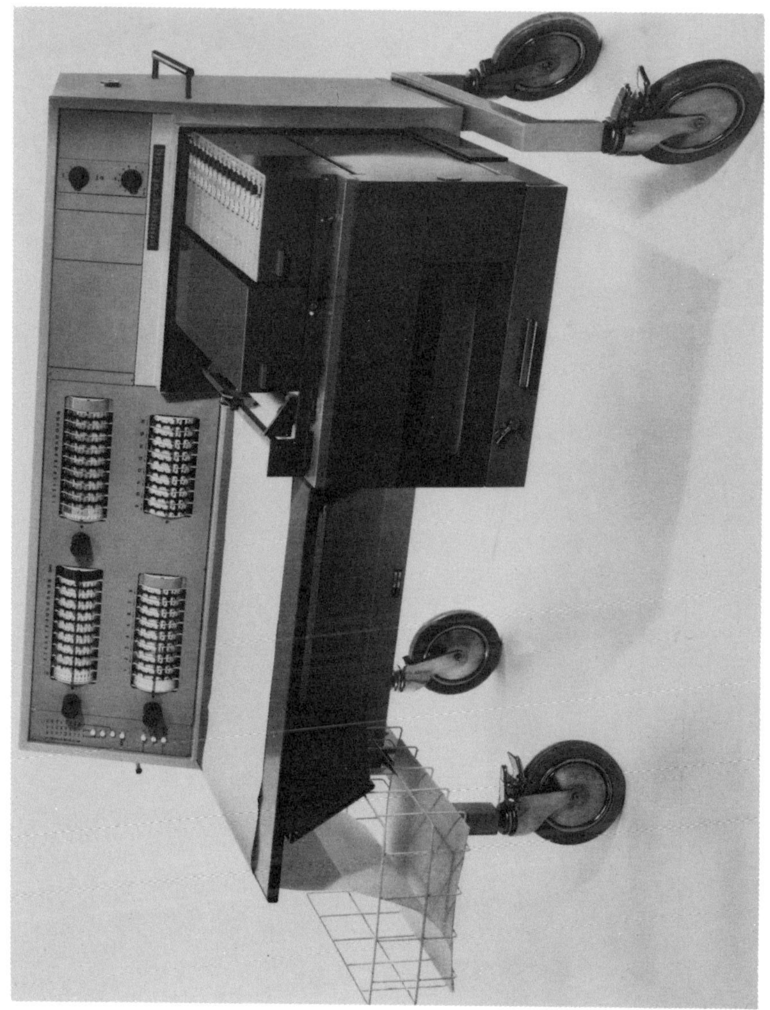

Figure 4-3. 16-channel electroencephalograph (courtesy of Siemens-Elema AB)

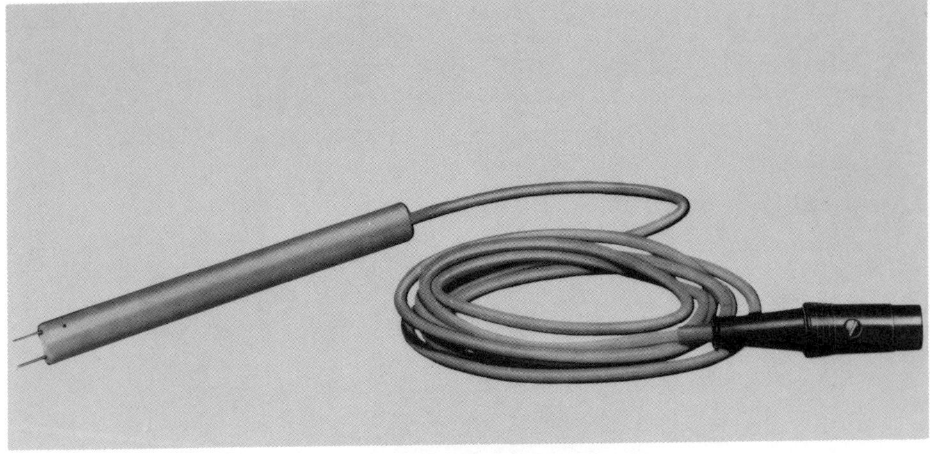

(a) Stimulation electrode

(b) Stimulation control unit

Figure 4-4. Electromyographic stimulation equipment (courtesy of DISA Elektronik A/S)

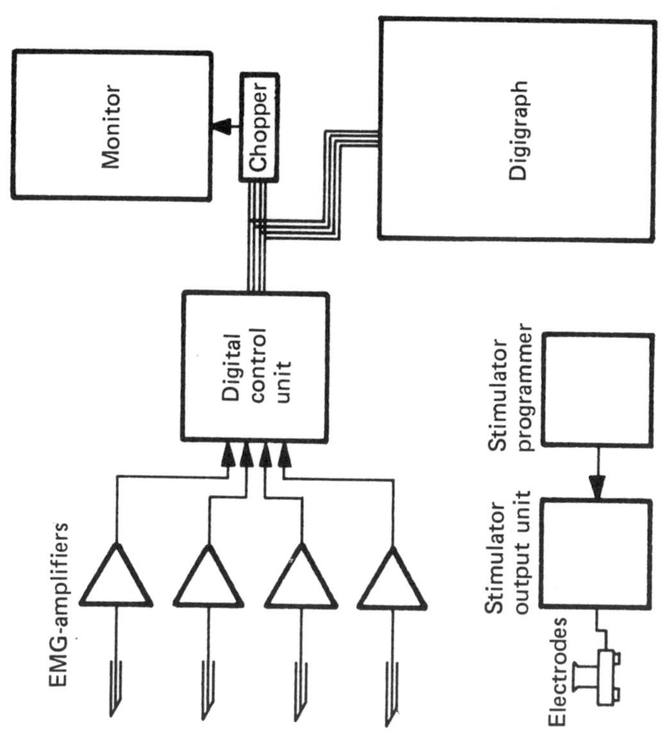

(a) EMG equipment console (digital)

(b) Simplified functional block diagram

Figure 4-5. Electromyograph (EMG) equipment (courtesy of DISA Elektronik A/S)

4.4 EXTERNAL SENSORY RESPONSE MEASUREMENTS

Direct measurements of the action and response of sensory organs have, so far, been limited to the eye. Additionally, semidirect measurements of the response of the ear and its associated brain-input nervous system have been in use for some time, in the form of *audiometry*. In its basic and most commonly used form, audiometry involves applying sound, at known amplitudes and frequencies, to the ear, such as by means of headphones worn by the subject. The audible stimulus is applied from an audio oscillator. This method is not a direct measurement method because the subject must stable his response (e.g., can hear, cannot hear) to the person administering the test, either verbally or by other consciously-expressed means. Nevertheless, audiometry is a useful tool for determining the overall threshold of hearing on a frequency-amplitude basis.

Direct measurements on the eye are of two types, the electroretinogram and the electro-oculogram. Additionally, indirect measurements, using positional information from a beam reflected by the eye, have been used to determine eye motion, including its viewing direction.

The graphic recording of changes in potentials generated by movements of the eyeball is an *electro-oculogram*. The potentials are sensed by electrodes placed on the skin surface close to the eye. One electrode pair is placed on the outer corners of the eye (the corner closest to the nose and the corner closest to the ear) to detect horizontal eyeball movement. Vertical eyeball motion is detected by two electrodes placed immediately below and immediately above the eye, respectively. Signals are within the overall range of 10 to 4000 μV but are usually in the nominal range 50 to 3500 μV. The electrodes are placed equidistantly from the pupil.

A calibration is first obtained with the eyeball in its center position. This requires a zero-suppression capability in the amplification/display equipment. Since motion can be quite rapid, a frequency response capability of 0 to 125 Hz is also required. Positional information is obtained by dc coupling, whereas relative movement is determinable by using ac coupling. An ungrounded bipolar electrode lead system is used for each electrode pair. The signal conditioner for an electro-oculogram (EOG) should have an input impedance of several megohms and a common mode rejection of at least 80 dB.

An *electroretinogram* (ERG) is a display of summed receptor potentials generated by the rods and cones in the retina in response to variations in incident illumination. The electrodes used to detect these potentials consist of a wire electrode, typically silver/silver chloride which is in electrical contact with a saline solution contained between the cornea and a contact lens placed over the cornea. A plate-type electrode placed on the forehead acts as reference (indifferent) electrode. Signal characteristics are amplitudes of less than 1000 μV and frequencies between 0 and 20 Hz.

5. Respiratory System Measurements

Diagnostic measurements of respiratory-system parameters, primarily those described in 2.4, can be categorized generally as follows:

(a) measurement of chest expansion and contraction;

(b) measurement of thoracic-cavity impedance;

(c) measurement of air flow during inspiration and expiration.

Abbreviations commonly used for functions related to respiratory system measurements (see 2.4) are shown in Table 5-1.

Chest-movement measurements show the physical motion of the thoracic cavity, as monitored by straps or tubes applied around the chest, during inspiration and expiration. The expansion and contraction of the strap or tube is detected by a sensor. Elastic straps equipped with either strain gages or displacement transducers have been used for this purpose, as well as a resistive device consisting of a rubber tube, filled with a copper-sulphate solution and provided with copper electrodes at each end. Since such devices do not constrain the subject (i.e., there is no need to breathe through a mouthpiece) they are useful in long-term monitoring of chest movements.

A similar advantage exists in the case of the *impedance pneumograph,* a sensing system consisting of two electrodes, an oscillator which provides an excitation current across the electrodes, and an amplifier which senses the changes in impedance across the electrodes. The electrodes are attached at the midaxillary line, usually between the 5th and 6th ribs, below each of the armpits. The impedance changes cause an output signal from the signal conditioner/amplifier from which changes in the respiratory volume can be determined. One instrument uses this arrangement as part of a cardiopulmonary measurement system. The same two electrodes are used as input to an EKG signal conditioner which is designed so as to reject the changes in the high-frequency (about 40 to 60 kHz) excitation voltage applied across the two electrodes for impedance measurements.

Table 5-1: Abbreviations Used in Respiratory-System Measurements

ERV – expiratory reserve volume

FEV – forced expiratory volume

FEV_t – forced expiratory volume in t seconds (timed vital capacity)

FEF – forced expiratory flow

FIC – forced inspiratory capacity

FRC – functional residual capacity

FVC – forced vital capacity

IC – inspiratory capacity

IRV – inspiratory reserve volume

IVC – inspiratory vital capacity

MVV – maximum voluntary ventilation

PEFR – peak expiratory flow rate

RV – residual volume

TLC – total lung capacity

VC – vital capacity

V_t – stroke volume

The respiratory-system sensing devices which are not most frequently used are those responding directly to air flow. Most of these are related to, or are special versions of, various types of commercial flowmeters. These devices have an output proportional to flow rate and this output can be either displayed directly or totalized for displays of total flow, under a given set of conditions, from which such pulmonary functions as lung volume can then be determined. Such devices are known as *spirometers* or *respirometers.*

Before describing some typical electronic respirometers, one device should be mentioned which is usually associated with a direct mechanical readout, but which indicates lung volume changes directly. This spirometer consists of an air-filled bell, inverted over a vessel filled with water, so that the bell moves up and down with changes in air volume while the water acts as a seal to the air. The bell is vented to a breathing tube, equipped with a mouthpiece at its other end. As the patient breaths through the breathing tube the up and down motions of the bell are recorded, usually by a pen riding on a cylindrical strip chart recorder *(kymograph)*. The bell is counterbalanced with a weight-and-pulley arrangement so that breathing (and its dispay) are not affected by the mass of the bell.

Since electronic respiratory-air-flow sensors are, essentially, gas flowmeters their sensing and transduction principles are mutually related. A cantilevered beam, to which strain gages are attached, and which is located within a breathing tube, has been used for bidirectional respiratory flow measurement (the polarity of the strain-gage-bridge output indicates flow direction). Thermal flowmeters, employing a heated platinum-wire or thermistor-type sensor within a breathing tube, are used for air-flow measurements but cannot indicate flow direction. The air-flow rate causes a proportional amount of cooling of the thermal sensor which is reflected by a proportional change in its resistance. These resistance changes are then converted to changes in voltage or current.

A turbine-type air-flow sensor is shown in Figure 5-1. The disposable flowmeter element consists of a plastic tube (whose end is the mouthpiece) containing a turbine-type impeller supported by bearings and containing a small through-hole. The angular speed of the impeller varies with air-flow rate. This disposable element is then inserted in a holder containing a light source and two phototransistors as well as an amplifier. As the impeller rotates it chops the light reaching the phototransistors. Their output is in terms of pulse-frequency which increases with increasing flow rate. The phototransistors are so spaced that a comparison of the phase of their output pulses also permits determination of flow direction. The purpose of the through-hole in the impeller is to alternatingly pass and block the light beam. The complete sensor is available as part of a (patented) system which includes a fiber-optics display matrix permitting direct film recording.

An electronic expirometer, also employing the turbine sensing element for flow transduction, is illustrated in Figure 5-2. A disposable mouthpiece mounts to the sensing assembly which contains a finned turbine element designed to rotate, in jewel bearings, at very high angular speeds (over 100 000 r/min at peak-flow exhalation). The display unit provides compensation for ambient temperature by means of a temperature-range selector switch. The unit can display FEV, FEV_t, and peak flow.

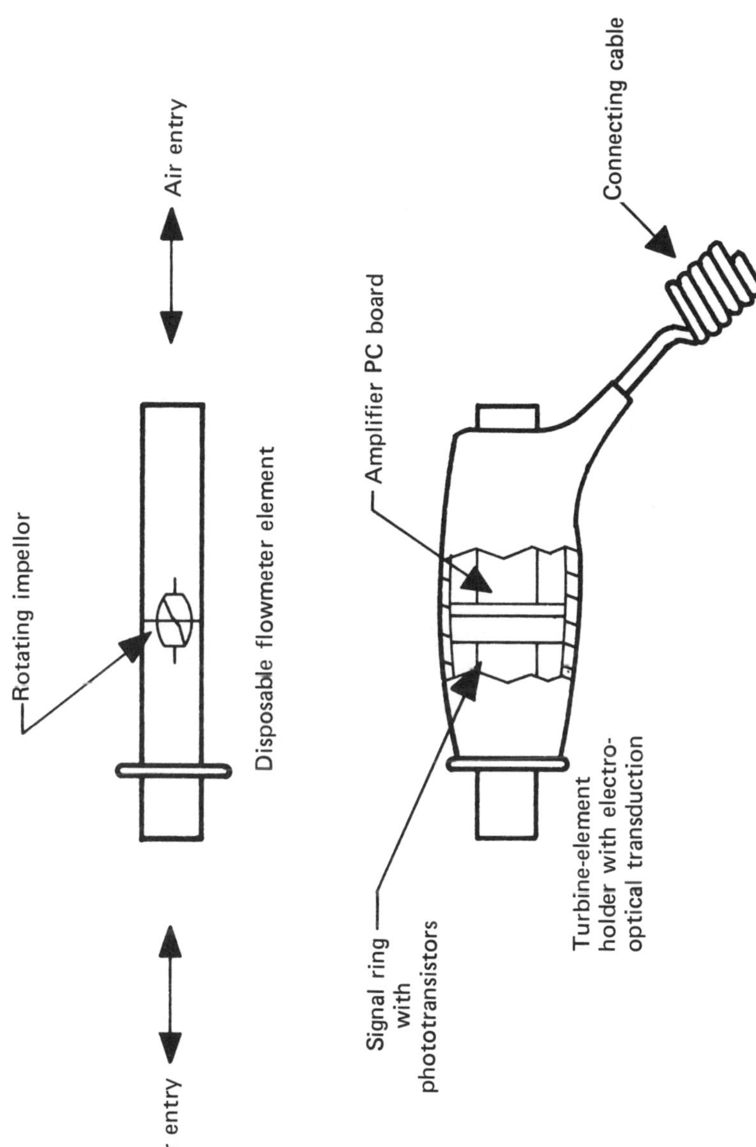

Figure 5-1. Turbine-type respiratory flow sensor (courtesy of Marion Laboratories, Inc.)

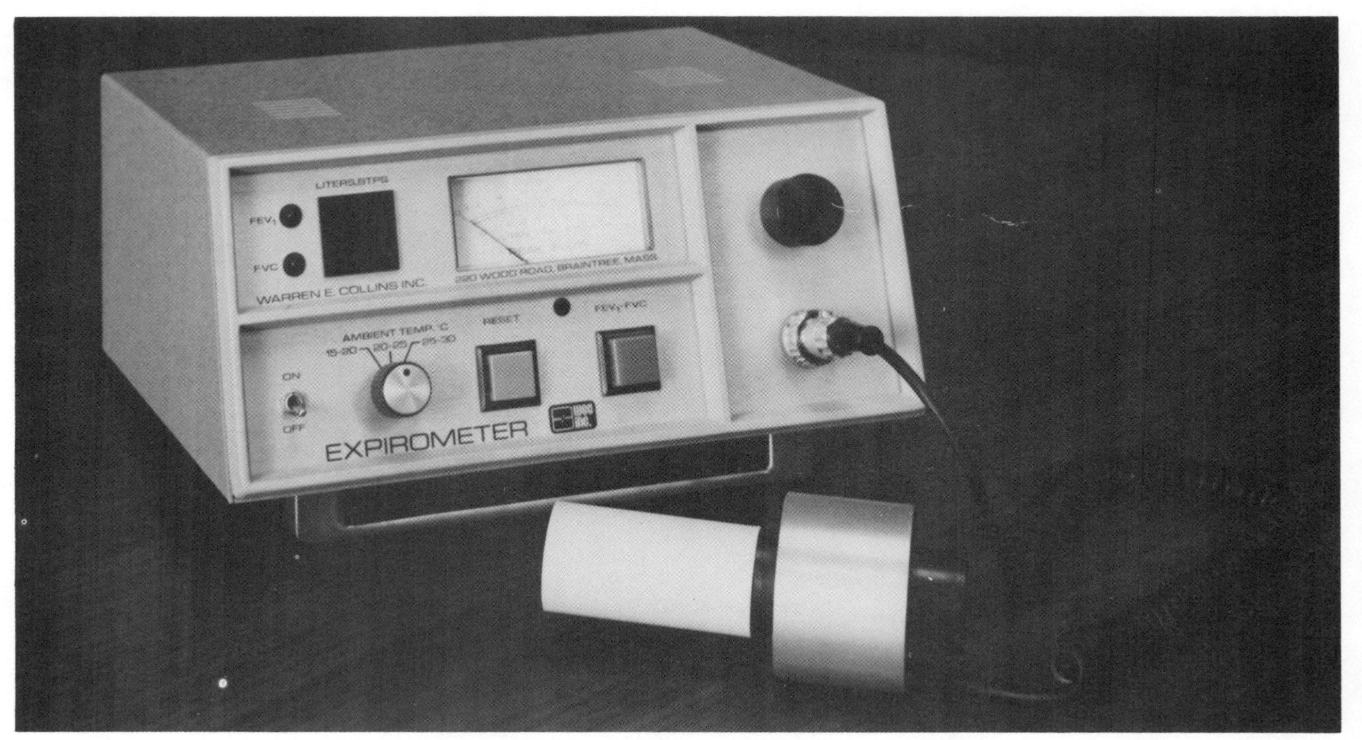

Figure 5-2. Turbine-type respiratory flow measurement system (courtesy of Warren E. Collins, Inc.)

Several types of respirometers employ a differential-pressure sensing principle. The air flow causes a proportional pressure difference across a restriction. The differential pressure appears across two pressure ports, one on either side of the restriction. Flexible tubing from the ports then leads to a differential-pressure transducer whose output signal varies with air-flow rate. Some of these sensors use only one port and one section of flexible tubing connected to a gage-pressure transducer, with the ambient pressure used as reference pressure.

One type of differential-pressure flow sensor is the variable-orifice type, illustrated in Figure 5-3. The variable orifice is a plastic membrane, sectioned into flaps (Figure 5-3b). The orifice adapts itself to variations in flow. It is closed at zero flow and opens to about 80% of the cross-section of the duct at maximum flow. The shape of the orifice tends to optimize the linearity of the characteristic curve (see Figure 5-3d) for bidirectional flow at varying velocities. The orifice material was chosen for maintaining its elasticity over a large number of measuring cycles and over a temperature range extending up to 70°C. The mouthpiece (see Figure 5-3c) is made of steam-resistant silicon rubber. Plug-in tubulation connects the pressure ports to a differential-pressure transducer in the display unit.

Other differential-pressure respiratory-flow sensors employ fixed-shape flow-restrictor systems to obtain a pressure difference proportional to flow rate. Mesh or grid assemblies and capillary systems have been used as flow restrictors. The sensor shown in Figure 5-4 uses a laminated element to obtain the differential-pressure output. The laminations provide a large number of parallel, rectangular, thin flow channels (Figure 5-4b). The differential pressure appears across the two pressure ports connected, through flexible tubing, to a differential-pressure transducer. An optional sputum trap and filter can be inserted between the mouthpiece and handpiece shown in Figure 5-4c. A comparison of the characteristic curve (Figure 5-4d) with that of the variable-orifice sensor shows that the sensitivity of the laminated type, in terms of differential pressure, is lower (0.15 mbar per ℓ/s as compared to 0.75 mbar per ℓ/s); however, the curve is more linear, particularly in the 0-10 ℓ/s region.

The back pressure built up behind a flow-restricting screen during exhalation can be measured as gage pressure (increasing with increasing flow rate), using a gage-pressure transducer, since both the flow sensor and the transducer are referenced to the ambient atmosphere. A sensing system employing this principle is shown in Figure 5-5. Flexible tubing from the flow sensor to the transducer (located in the readout unit) carries the back pressure generated in the sensor. The ancillary equipment shown consists of a digital printer and a strip-chart re-

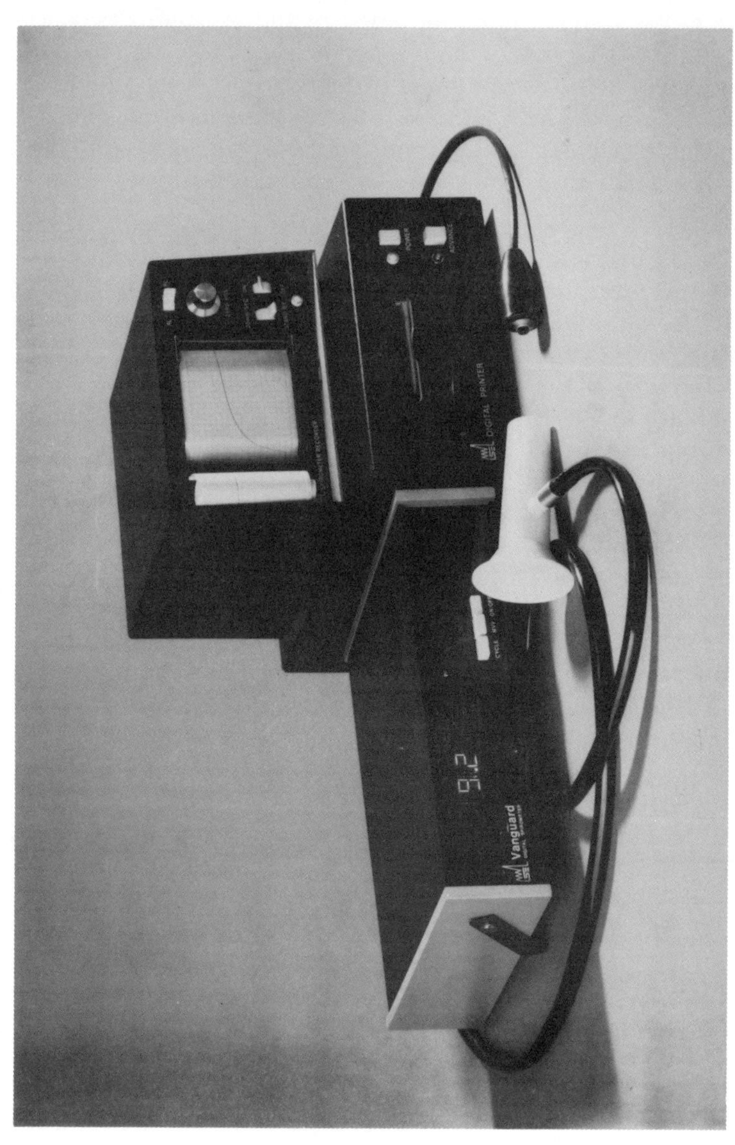

Figure 5-5. Differential-pressure type respiratory-flow sensing system (pulmonary function system) using screen in sensing element to build up a back pressure which varies with flow (courtesy of Life Support Equipment Corp.)

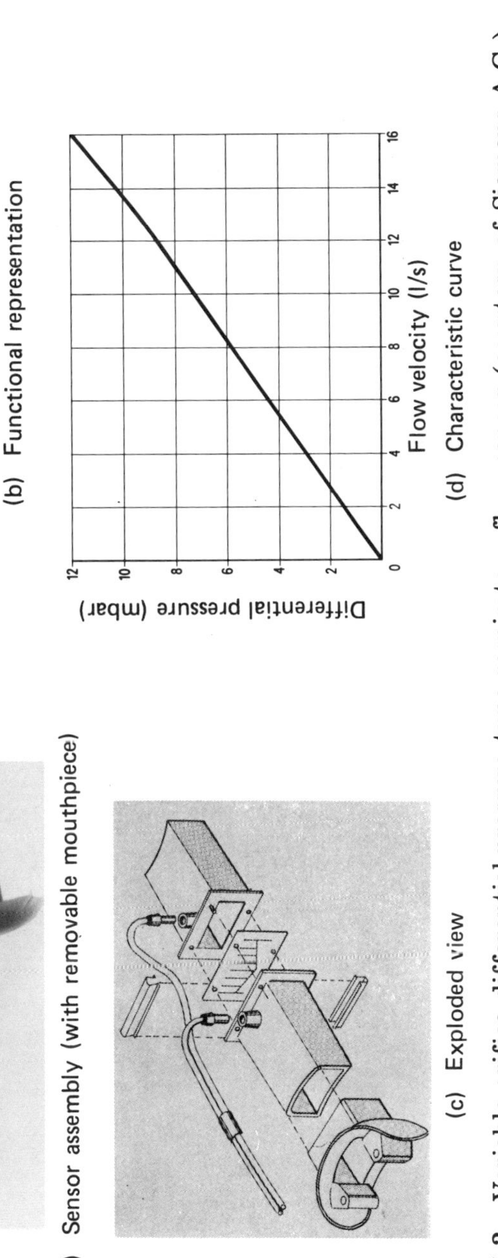

(a) Sensor assembly (with removable mouthpiece)

(b) Functional representation

(c) Exploded view

(d) Characteristic curve

Figure 5-3. Variable-orifice differential-pressure type respiratory-flow sensor (courtesy of Siemens A.G.)

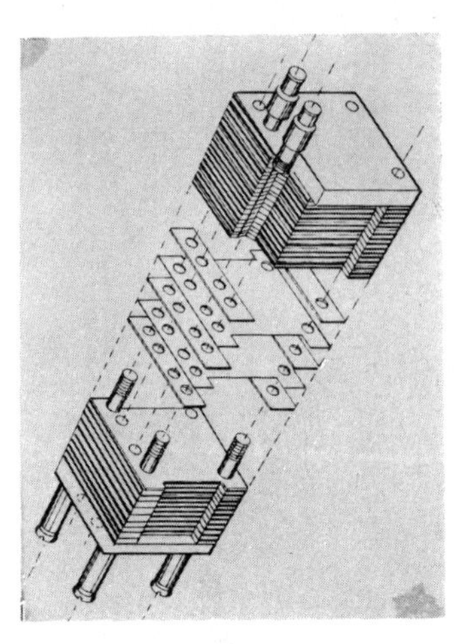

(a) Sensor assembly with connecting tubing

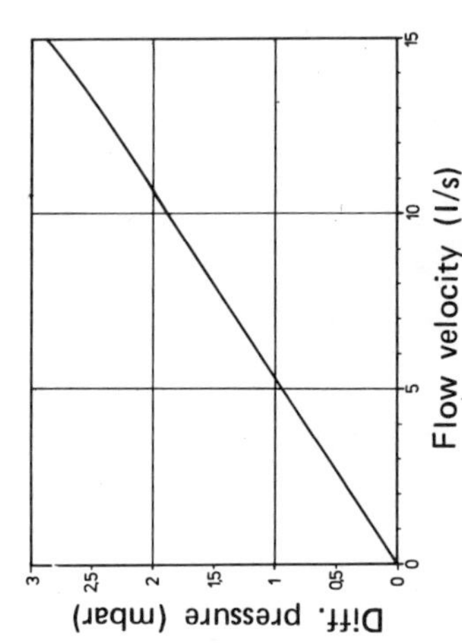

(b) Internal construction of sensing element

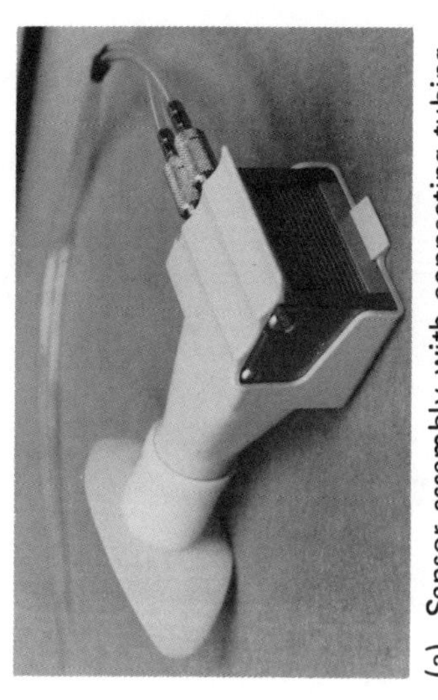

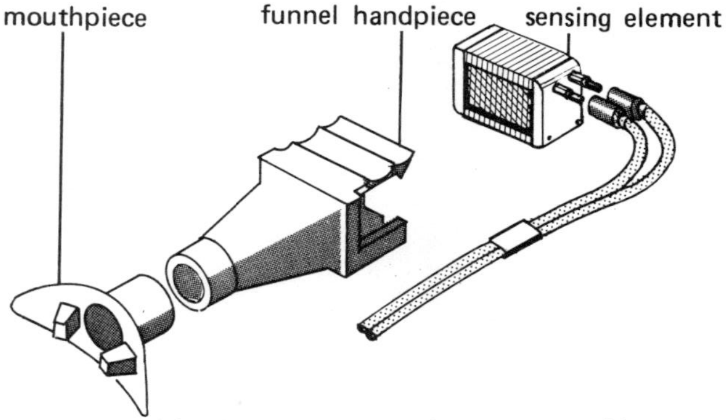

(c) Major elements of sensor assembly

Diff. pressure (mbar)
3
2,5
2
1,5
1
0,5
0
0
5
10
15
Flow velocity (l/s)

(d) Characteristic curve

Figure 5-4. Differential-pressure type respiratory-flow sensor with laminated sensing element (courtesy of Siemens A.G.)

corder. The disposable sensor offers a flow resistance equal to 0.31 cm H_2O per liter-second. As in most such systems, readout is provided for flow (flow rate, ℓ/s) as well as volume (in liters, by integrating flow). Besides peak flow, readout is provided for FVC, VC, FEV_t as well as FEV at 0.5, 1.0 and 3.0 s, and % FEV_t as well as % FEV at 0.5, 1.0 and 3.0 s.

6. Skin Measurements

The most commonly measured skin parameters are those related to perspiration, the *basal skin resistance* (BSR) and the *galvanic skin response* (GSR). They are usually measured at the palms of the hands, sometimes at the soles of the feet. The most commonly used sensing points are at the center of the palm, for the active electrode, and the wrist, for the neutral or reference electrode. A basic diagram for the measurements is shown in Figure 6-1.

With the switch closed a dc measurement of the static value of skin resistance (BSR) is obtained. With the switch in the open position the signal into the amplifier is ac-coupled and the dynamic variations of skin resistance (GSR) are measured. The combination of the capacitor and amplifier input resistor forms an RC network typically selected to provide a time constant of about 4 seconds. GSR measuring equipment is usually provided with a capability to balance out the BSR so that well-amplified displays of GSR can be obtained.

GSR is normally the measurement of greatest interest. It is really one of a category of psychophysiological measurements. Psychological stimuli are used to obtain emotional responses from the subject. These responses are then made visible in the form of resistance-variation displays. An example of GSR equipment is the "lie detector" used in police sciences.

A skin parameter whose variations are sometimes measured, also as functions of emotional responses, is *skin potential,* typically sensed between palm and forearm and in the range of about 50 mV.

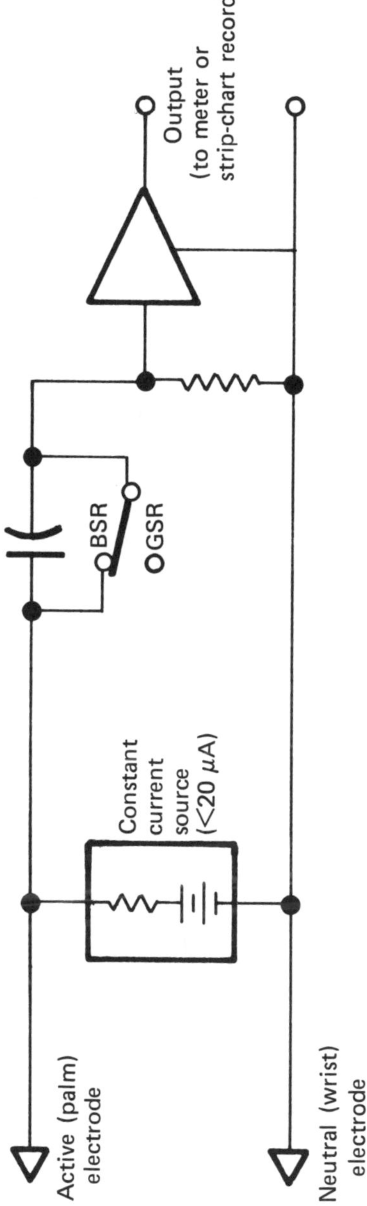

Figure 6-1. Simplified diagram of BSR and GSR measuring system

7. Other Biomedical Measurement Instrumentation

In addition to the sensing devices for specific systems of the body, described above, there are a considerable number of sensing systems and combinations of systems which had their origin in nonbiomedical applications but are now used in the biomedical field. Some of the more significant of these are briefly covered in this section.

7.1 TEMPERATURE

Small mercury-in-glass thermometers are still widely used for routine oral and rectal temperature measurement. However, electronic thermometers, which are now available at very reasonable cost, have been replacing the glass thermometers increasingly, not only in clinics and hospitals, but in doctors' offices and even in private homes. Virtually all electronic thermometers use a thermistor as sensor (see Chapter 1), since thermistors usable for such applications have a fairly high temperature coefficient of resistance (around 4% per degree Celsius).

The thermistor, usually of the small bead type, is connected into a bridge or voltage divider circuit so that its resistance changes can be converted to voltage changes which, sometimes electronically amplified, can be displayed on a meter (or other display device). The meter scale is usually limited to the medical range of interest, e.g., 35 to 40°C (95 to 104°F) for routine applications.

The primary advantage of the electronic thermometer is the time required to obtain an accurate temperature measurement: around 20 seconds, as compared to between 3 and 5 minutes for mercury-in-glass thermometers. Another advantage is the absence of mercury which presents a hazard if the thermometer breaks.

The sensing portions of electronic thermometers are sometimes constructed as probes, for rectal use, sometimes of special configuration for surface, internal or implantable use. The most common configuration, however, is a small bead thermistor at the end of a thin plastic rod, for placement under the tongue near the sublingual artery. Thin throw-away plastic sleeves are available for placing around the probe tip so that sterilization is not required. Alternatively, the probes can be cleaned in a sterilizing liquid.

The compact, hand-held electronic thermometers are battery-operated and include provisions for calibration, to allow for variations in battery voltage or minor changes in circuit component characteristics. The slight nonlinearity of the thermistor in the medical measuring range is compensated either by linearizing circuits or by making the meter scale appropriately nonlinear. Multipatient monitoring has been installed in hospital wards, using a sensing probe at each bedside and a central probe-selection and display unit. Temperature displays can be in digital or analog form. The temperature-dependent signal can also be fed to a strip-chart recorder or a computer or digital printer.

Thermography is a method for obtaining a graphic display of skin temperature variations over large portions of the body. An infrared imaging system is employed for this purpose, using a scanning radiation pyrometer (infrared thermometer) as sensor.

7.2 DISPLACEMENT, FORCE AND ACCELERATION

Sensors for these solid-mechanics measurands (see Section 1.4) are also used for some biomedical applications. Force and acceleration measurements are made during stress-testing (while the subject is performing specific exercises), during conditions where abnormal forces or accelerations are acting on the subject, to determine muscular force, and to determine forces that an artificial limb or bone-joint must be able to withstand.

Displacement sensors have been used to determine the mobility of limbs and the rotatability of joints. A number of other applications (pulse sensing, chest expansion measurement, etc.) have been covered in previous chapters of this book. An example of the variety of additional applications is an eyeball-motion sensor of the infrared photoelectric type which is small enough to allow attachment to an eyeglass frame. Since the iris absorbs infrared energy whereas the surrounding area of the eyeball reflects it, eyeball motion can be detected by the light-source and light-sensor combination without affecting the subject's visual activity and without sensing effects of ambient light in the visible spectrum.

Uses of motion, acceleration and force sensors are also found in various medical research activities as well as in developmental tests on structures and vehicles where effects of noise, accelerations and decelerations on the comfort and safety of human occupants are to be determined. An example of the latter is the use of life-size dummies, heavily instrumented with accelerometers and strain gages, during automotive crash testing. An understanding of the different types of motion, strain, force, and acceleration sensors, and of their static and dynamic characteristics, is necessary for the design, implementation and use of such solid-mechanical measurement systems.

7.3 RADIOLOGICAL, NUCLEAR AND ULTRASONIC INSTRUMENTATION

Sources of, and sensing devices for, electromagnetic radiation in the nuclear-radiation, X-ray, and ultrasonic bands of wavelengths are used in medicine for diagnostic purposes as well as for therapy (therapy instrumentation is not a subject of this book).

Sensors (and sources) of *ultrasound* are generally piezoelectric, sometimes magnetostrictive in transduction principle. The widely-used piezoelectric transducers are often used as source as well as sensor. When a piezoelectric crystal is excited by a source of high-frequency alternating current it undergoes mechanical deformations; these dimensional fluctuations are coupled to a flexible diaphragm from which ultrasound is then emitted (at the frequency of the excitation voltage). When this type of device is not excited, the diaphragm will deflect (very slightly) with ultrasound received, couple these deflections mechanically to the piezoelectric crystal and then cause an electrical signal to be produced by the crystal as it undergoes mechanical deformations. Electronic circuitry is available to alternatingly apply a very brief burst of excitation voltage to the crystal and then revert to a "listening" mode.

The source usually employed in *X-ray* diagnostics is an X-ray lamp. X-rays are emitted by a metallic target in the lamp as the target is struck by a high-energy electron beam. Associated equipment permits adjustment of the cathode current (and amount of electron emission) and of the high voltage between cathode and anode, as well as of the duration of X-ray emission (dosage). The X-rays are usually detected by a photographic film with an X-ray sensitive emulsion; however, in other applications they can be detected by gas-filled ionization chambers or semiconductor detectors.

Nuclear radiation sources are usually radioactive isotopes of certain elements. The source may be external or internal to the body. Detectors are selected on the basis of type of radiation (gamma-, beta- or alpha-ray) to be sensed as well as the energy of the radiation (energy is inversely proportional to wavelength). They include gas-filled ionization chambers and various types of semiconductor detectors that respond directly to the radiation incident upon their window or sensitive area; scintillation detectors in which the radiation interacts with a scintillator material which then emits photons that can be detected by a light sensor (photodetector); and neutron detectors in which the neutrons first interact with a conversion material.

X-rays and ultrasonics are commonly used in those diagnostic procedures which result in producing an image of a portion of the body. Nuclear radiation is used primarily in diagnostic procedures employed to determine the functions and characteristics of organs, of the brain, of blood vessels and of some elements of the skeletal and nervous systems. X-ray images are usually created on a special photographic film. After developing, the film can be analyzed visually. The image will have been created by variations in the attenuation of the X-rays as they pass through the region of the body that is to be analyzed.

The film can also be scanned by a digital imaging system. This permits image enhancement and other forms of image processing that allow significant improvements in the display of detailed features of interest. Ultrasonic and nuclear-radiation images are provided by the output of sensors. One or more sensors are used in scanning equipment in which the sensors are moved, in accurately-controlled rectilinear motion, over the area of interest. By repeating the horizontal scan sequentially over vertically adjacent swaths, an image is created line-by-line.

Alternatively, multisensor arrays are being developed so that a complete image can be obtained at one time, without scanning. Rapid electronic scanning of the outputs of the sensor elements in the array then leads to the reconstruction of an image on a cathode-ray tube, on hard copy, or both. The sensor outputs can also be digitized and computer-processed for more specific analyses.

Nuclear radiation is used for imaging as well as for tracing; the latter involves the use of only one radiation detector that is placed or held over the region being examined. In both types of procedures, a specific radioisotope *(radionuclide)* is selected, and injected into a blood vessel, an organ, or a bone, where the radionuclide is absorbed or reacted to by that element of the body. The radionuclide is selected so as to optimize its absorption, or a reaction, in a specific part of the body. In brain scanning, for example, the brain's uptake of ytterbium-169, indium-113m, or technetium-99m is determined by scintillation-counter imaging of the brain; phenomena such as brain lesions can then be detected from the image.

Tomography is a form of X-ray or (now increasingly) ultrasonic imaging by which an image of a single selected plane within a portion of the body is produced, with the outline of structures in other planes essentially eliminated. By sequentially imaging increasingly deeper-lying planes, a three-dimensional picture can be reconstructed. Computers are used to calculate the appropriate spatial coordinates and to control scan and penetration depth as well as to process the sensor-provided information and display the desired type of picture. These functions are all provided by the *computerized axial tomography* (CAT) *scanner,* which is used for such imaging of, e.g., the brain and the thoracic region.

7.4 PATIENT MONITORING SYSTEMS

Combinations of sensing devices for different biomedical measurements, employed simultaneously, with signal processing and display equipment, are now commonly used as patient monitoring systems in hospitals, particularly in intensive-care and coronary units. Such systems permit the continuous surveillance of all essential physiological data of the patient. Typical measurements made and displayed simultaneously are respiration rate, heart rate, temperature, blood pressure, and heart potentials (EKG). The data from several bed stations are often displayed on one central console. Either hardwiring or radio telemetry is used for signal transmission to the display station.

Many patient monitoring systems now in use employ computers for signal storage and, especially, for signal processing. Relatively simple programs allow checking whether the signal is within preset limits, correlation of different signals and detection of changes in signals from previous values and the magnitude of such changes, all of which can be used to activate a visual or audible alarm. The data base of the computer base can also often be accessed not only from the central display station but also from other terminals located in different parts of the hospital (e.g., surgery, clinical laboratory, etc.).

A computer-assisted patient monitoring system is illustrated in Figure 7-1, with measurements such as heart rate, respiration rate, temperature, and blood pressure displayed on the monitoring console. This system permits data storage for periods up to 16 days.

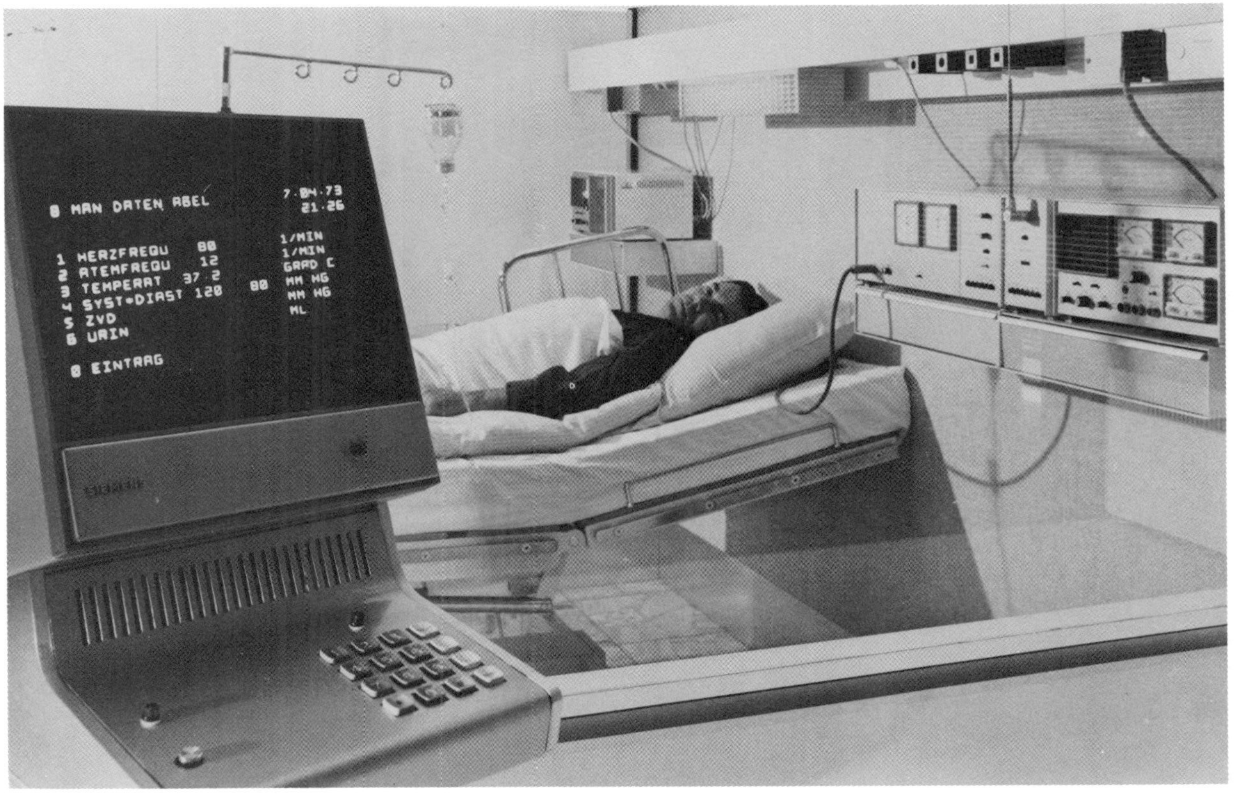

Figure 7-1. Computer-assisted patient monitoring system (courtesy of Siemens A.G.)

8. Electrical Safety Considerations

Adherence to good electrical safety practices is of extreme importance in the use of biomedical electronic equipment. Improper grounding or isolation and lack of safeguards against excessive currents (e.g., leakage currents) that could be encountered can be hazardous not only to patients but also to nurses and technicians and even to bystanders.

It is interesting to note that adherence to good electrical safety practices has been strongly emphasized in the aerospace industry (the origin of many biomedical electronics concepts), but this awareness has just recently (middle 1970s) penetrated to designers and users of biomedical electronics equipment. Most of the electrical safety practices, applicable to the biomedical field, are contained in government or industry standards, such as (in the U.S.A.) the National Electrical Code (NEC) and standards issued by the National Fire Protection Agency (NFPA). Additional standards have been issued by medical groups and hospital associations.

Of paramount importance is an equipotential electrical environment in any area where electronic equipment is used for patient monitoring (or treatment). This means that all accessible metallic surfaces that might contact ground or a source of leakage current must be connected by well-attached wires to a common ground terminal which, in turn, is well grounded as part of the overall electrical system. This includes the cabinets of electrically-powered medical equipment as well as such often-overlooked items as plumbing (faucets, pipes, etc.) and heating devices (such as hot-water radiators) and oxygen or suction outlets.

Only grounded electrical outlets are to be used for ac-operated equipment. The power cord of the equipment must include a properly-connected ground wire as part of the (three-wire) cord, and the grounding contact in the receptacle must firmly grip the ground pin of the plug.

All power-carrying wires must be of the proper gage. The use of battery-operated equipment eliminates potential hazards associated with the building's power system. When equipment is furnished with more than one external receptacle, the receptacles must be of different configuration (or pin arrangement, or key orientation) so that it is impossible to plug a cable into the wrong connector. Power-carrying cables should be as short as feasible to minimize their ohmic resistance and reduce leakage current.

Components within equipment should be selected (and tested) for good insulation resistance and proper breakdown-voltage rating. The points to which patient electrodes or wiring from patient-contacting sensors are connected to the equipment should preferably be electrically isolated from the other circuitry in the equipment. Connections, especially ground connections, should be tested periodically for their integrity, as well as for their proper insulation from other terminals.

Although standards and directives for maximum permissible currents that are caused to flow through a patient differ somewhat, it is generally accepted that currents between patient connections should be less than 50 microamperes rms ac, unless the current can pass directly through the myocardium in which case a limit of 10 microamperes applies. Leakage current from equipment that is not to be connected to patients must be limited to 500 microamperes. This is the threshold of perception, the smallest current that can be felt. Another current limit to remember is the "let-go" limit of 15-20 milliamperes rms; if above this limit, a person accidentally grasping two electrical connections and being subjected to electrical shock may find it impossible to release himself from the electrical contact.

Because of the need for electrical safety tests in the biomedical field, a number of manufacturers are offering safety analyzers for such purposes. The analyzer shown in Figure 8-1, for example, permits measurement of leakage current, including on patient-connection leads, of power line voltage, of power wire resistance, of ground resistance, and of balance of isolated power systems. Special terminals are provided for performing all recommended testing of electrocardiograph leads.

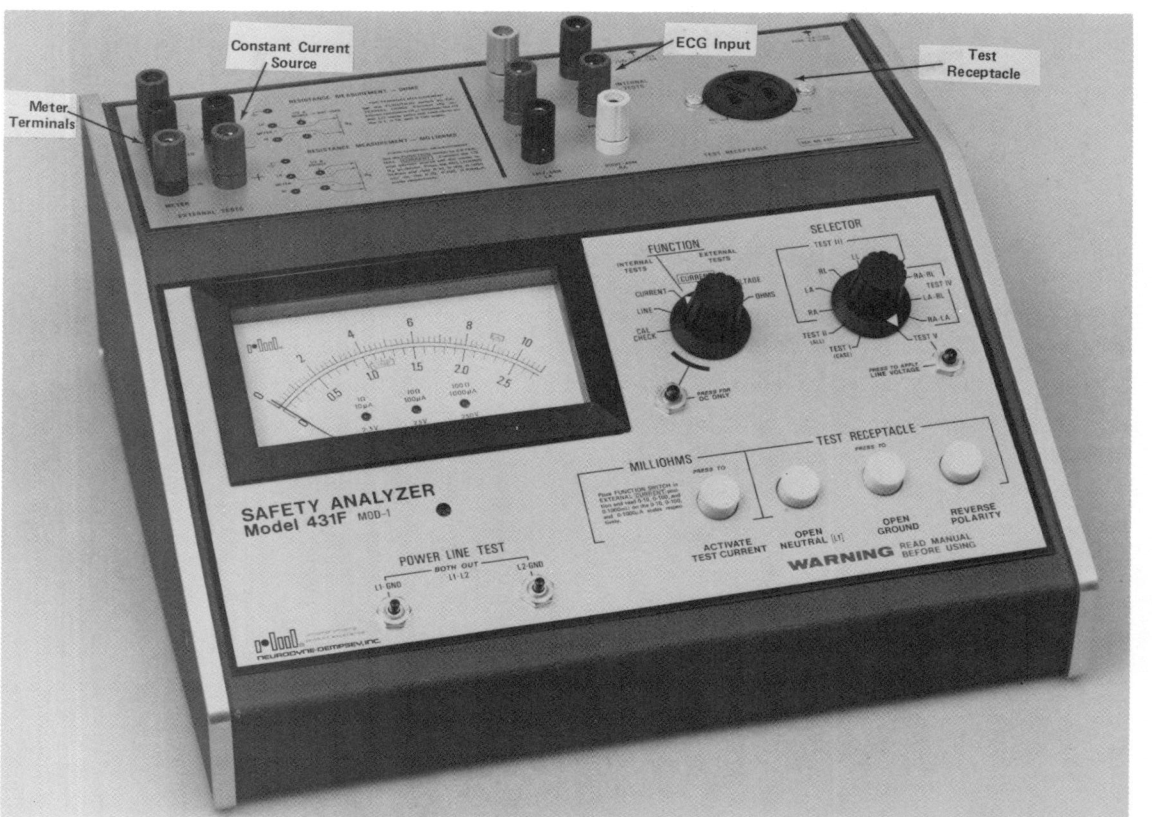

Figure 8-1. Safety analyzer for biomedical applications (courtesy of Neurodyne-Dempsey, Inc.)

9. Bibliography

9.1 GENERAL INSTRUMENTATION TOPICS

Liptak, B.G., *Instrument Engineers Handbook,* vol. II. Philadelphia, PA: Chilton Book Co., 1970.

Norton, H.N., *Trilingual Dictionary of Measurement Terms* (English-German-French). Lausanne, Switzerland: Scriptar, Ltd., 1970.

Peatman, J.B., *The Design of Digital Systems.* New York: McGraw-Hill Book Co., 1972.

Considine, D.M. (Ed.-in-Chief), *Process Instruments and Controls Handbook* (2nd ed.). New York: McGraw-Hill Book Co., 1973.

Willard, H.H., Merritt, L.L., Jr. and Dean, J.A., *Instrumental Methods of Analysis* (5th ed.). New York: D. Van Nostrand Co., 1974.

Fink, D.G. (Ed.-in-Chief), *Electronic Engineers Handbook.* New York: McGraw-Hill Book Co., 1975.

Handbook of Chemistry and Physics (55th ed.). Cleveland, OH: CRC Press (Chemical Rubber Co.), 1976.

Benedict, R.P., *Fundamentals of Temperature, Pressure and Flow Measurements* (2nd ed.). New York: John Wiley & Sons, 1977.

Morrison, R., *Grounding and Shielding Techniques in Instrumentation* (2nd ed.). New York: John Wiley & Sons, 1977.

Standards and Practices for Instrumentation (5th ed.). Research Triangle Park, NC: Instrument Society of America, 1977.

Norton, H.N., *Sensor and Analyzer Handbook.* Englewood Cliffs, NJ: Prentice-Hall, Inc., 1982.

9.2 BIOMEDICAL TOPICS

Traite, M., Welkowitz, W., Kilduff, J. and Purpuro, C. "Environmental Testing of Future Spacemen," *Electronics,* vol. 32, no. 42, 1959.

Kugler, J., *Electroencephalography in Hospital and General Consulting Practice.* New York: American Elsevier Publishing Co., 1964.

Smith, C.R. and Bickley, W.H., *The Measurement of Blood Pressure in the Human Body,* NASA SP-5006. Washington, DC: U.S. Government Printing Office, 1964.

McCaughan, D., "Comparison of an Electronic Blood Pressure Apparatus with a Mercury Manometer," *Journal for the Advancement of Medical Instrumentation,* Nov./Dec. 1966.

Venables, P.H. and Martin, I. (Eds.), *A Manual of Psycho-Physiological Methods.* New York: John Wiley & Sons, Inc., 1967.

Geddes, L.A. and Baker, L.E., *Principles of Applied Biomedical Instrumentation.* New York: John Wiley & Sons, Inc., 1968.

Mackay, R.S., *Bio-Medical Telemetry.* New York: John Wiley & Sons, Inc., 1968.

"Biomedical Measurements," Home Study Course 13, *Measurements & Data,* Jan./Feb. 1969, Pittsburgh, PA, 1969.

Katona, Z. and Bolváry, G., "Medico-Technical Problems in Blood Pressure Measurement and Their Solution by an Automatic Measuring Technique," Paper No. B-505, *Acta IMEKO,* 1970. Budapest, Hungary: IMEKO, 1970.

Quimby, E.H., Feitelberg, S. and Gross, W., *Radioactive Nuclides in Medicine and Biology: Basic Physics and Instrumentation.* Philadelphia, PA: Lea & Febiger, 1970.

Rasmussen, C.G., "A Catheter Thin-Film Probe for Measurement of Instantaneous Blood Flow," Paper No. B-506, *Acta IMEKO,* 1970. Budapest, Hungary: IMEKO, 1970.

Strong, P., *Biophysical Measurements.* Beaverton, OR: Tektronix, Inc., 1970.

Brown, J.H.V., Jacobs, J.E. and Stark, L., *Biomedical Engineering.* Philadelphia, PA: F.A. Davis Co., 1971.

Evans, W.F., *Anatomy and Physiology, the Basic Principles*. Englewood Cliffs, NJ: Prentice-Hall, Inc., 1971.

Guyton, A.C., *Textbook of Medical Physiology*. Philadelphia, PA: W.B. Saunders Co., 1971.

Manual for the Safe Use of Electricity in Hospitals, 76 BM. Boston, MA: National Fire Protection Association, 1971.

"Pulmonary-Respiratory Measurements," *Medical Electronics and Data,* Jan./Feb. 1972.

Wartak, J., *Phonocardiology*. Hagerstown, MD: Harper & Row, Medical Dept., 1972.

Watson, J. and Rogers, J.E., "Metal-on-Glass Microelectrodes," *Medical Electronics and Data,* Jan./Feb. 1972.

Brown, P.B., Maxwell, B.W. and Moraff, H., *Electronics for Neurobiologists*. Cambridge, MA: The MIT Press, 1973.

Fitzgerald, M.X., Smith, A.A. and Gaensler, E.A., "Evaluation of Electronic Spirometers," *The New England Journal of Medicine,* vol. 289, no. 24, Dec. 13, 1973.

Frenay, A.C., Sr., *Understanding Medical Terminology*. St. Louis, MO: The Catholic Hospital Association, 1973.

Ferris, C.D., *Introduction to Bioelectrodes*. New York: Plenum Publishing Corp., 1974.

Thomas, H.E., *Handbook of Biomedical Instrumentation and Measurement*. Reston, VA: Reston Publishing Co., 1974.

"Electrical Safety," *Home Study Courses No. 42/24 and 43/25*. Pittsburgh, PA: Measurement and Data Corp., 1973 and 1974.

Spooner, R.B. (Ed.), *Hospital Instrumentation Care and Servicing*. Pittsburgh, PA: Instrument Society of America, 1977.

Cromwell, L., Weibell, F.J., Pfeiffer, E.A. and Usselman, L.B., *Biomedical Instrumentation and Measurements* (2nd ed.). Englewood Cliffs, NJ: Prentice-Hall, Inc., 1980.

Weibell, F.J., *Biomedical Measurement Instrumentation* (Measurement Instrumentation Systems, vol. II, H.N. Norton, Ed.-in-Chief). New York: Garland STPM Press, 1982.

Index